EVSL2/NE3722

3.2.87

EVSL2/NE3722

3.2.87

SOIL STRUCTURE ASSESSMENT

Soil Structure Assessment

Edited by

W. BURKE
Kinsealy Research Centre, Dublin, Ireland

D. GABRIELS
Department of Soil Physics, State University of Ghent, Belgium

J. BOUMA
Soil Survey Institute, Wageningen, Netherlands

*Sponsored by the Commission of the European Communities,
Directorate-General for Agriculture, Coordination of Agricultural Research*

A.A. BALKEMA / ROTTERDAM / BOSTON / 1986

Published for the Commission of the European Communities,
Directorate-General Information Market & Innovation, Luxembourg

EUR 9497 EN

ISBN 90 6191 656 9

Published by A.A.Balkema, P.O.Box 1675, 3000 BR Rotterdam, Netherlands
Distributed in USA & Canada by: A.A.Balkema Publishers, P.O.Box 230, Accord, MA 02018
Printed in the Netherlands

Contents

Introduction

The concept of soil structure has many aspects. Soil structure, as used in this publication, refers to the physical constitution of soil material as expressed by the size, shape and arrangements of soil particles and voids and its associated properties. There is general agreement on the role of fundamental soil properties such as, for example, texture, pore size distribution, density, and plasticity which are important soil structure features. There are several accepted methods for determining such properties, and individual laboratories have adapted some of these methods in response to particular needs, depending on circumstances and requirements. Because soil structure is never static, methods which suit one set of circumstances may be unsuitable when applied elsewhere.

The measurement of soil structure is complex, and it is unlikely that attempts to measure it completely will ever be fully successful or that it will be possible to define a single, unique set of measurement procedures for all soils. Most workers compromise by measuring only that which they consider relevant to their particular purpose. In addition, some research workers make special, specific measurements that they regard to be important in relation the specific needs of their own projects. In the total context of soil structure studies this all adds up to a large number of measurements which are made in different ways.

The development towards uniformity and eventual standardisation of methods of soil structure determinations was already receiving attention at the 6th International Congress of Soil Science in Paris in 1956. A start in co-operation was made at the International Symposium on Soil Structure in Ghent (Belgium) in 1958, when a West-European Committee was set up. This committee reported to the 7th International Society of Soil Science Congress in Madison in 1960. Subsequently, regional working groups were established world wide. The West-European group remained active for several years, but despite all efforts of the group, it turned out to be impossible to recommend a series of standard methods. However, a series of representative methods was assembled from those likely to be suitable as reference methods and presented in a large volume entitled: "West-European Methods for Soil Structure Determination" (1967). The volume listed and described in varying details the methods in use in most West European laboratories. In 1965 the American Society of Agronomy published the well-known Agronomy 9 book "Methods of Soil Analysis", among others containing a detailed listing of methods to be used for soil structure assessment. A revised version is to appear in 1986. The East European countries published their Method Book in 1968.

Definition of reference methods became urgent when the Commission of the European Economic Communities (EEC) became active in soil structure research in its Land Use and Management Programme as related to Land Resource Development and Conservation. Accordingly, workshops on "Soil Structure Methodology" were organised to consider how the methods could as far as possible be "standardised" for use in future EEC co-ordinated "land use" investigations. At the first meeting it was agreed to consider three levels of standardisation: (a) Methods which in principle ought to be standardised, (b) Methods where variations exist because the underlying principles are difficult to define precisely; for these, reference methods are needed so that the variations can be related to each other, (c) Methods where flexibility is needed. For this group it was considered that standardisation is not only impossible but at present it is not even useful.

It was soon apparent and a fear was expressed that standardisation could become synonymous with inflexibility. There even was a warning for creating an illusory standardisation becoming a purpose in itself. Methods, for which standardisation is not possible at present, are vital for some objectives. Many methods are still evolving and standardisation might possibly impose too many restrictions and would probably curtail natural evolution. Following extended discussions, agreement was reached on some course of action towards unification in methodology. **Reference** methods rather than **standard** methods were to be recommended for workers carrying out EEC funded research projects. These reference methods, which are widely used, can either be used directly by researchers or can be used to calibrate other methods which individual researchers may be using. The use of such reference methods would enable more meaningful comparisons to be made between results obtained in different laboratories or countries.

The workshop invited individual members to contribute chapters on the topics being selected. Detailed descriptions of experimental procedures were not to be provided, as they are available elsewhere. In the book, reference methods are briefly described and sources which can provide specific information are cited. Special emphasis is given to operational aspects of the methodology. This represents a unique feature of this particular publication. Usually, methods are described in technical, scientific terms only. Operational aspects are, however, crucial for practical applications. Emphasis in the book is therefore placed on such aspects as cost, complexity, duration, accuracy and applicability in various soils and landscapes. In addition to the usual monodisciplinary treatment, interdisciplinary approaches to the subject have also been selected. Soil structure assessment has so far been widely covered from either an exclusive soil-physical or a morphological soil-survey point of view. A modest attempt is made in this book to indicate areas where a combined approach migh be profitable.

Although the book is the result of contributions by all members attending the workshops, certain individuals were given the daunting task of presenting the contributions in a uniform format. The site and soil description was prepared by D. Gabriels (Belgium). J. Bouma (The Netherlands) was responsible for Chapter 2: "Sampling for soil structure characterisation". A. Newman (United Kingdom) prepared part of the draft for the chapter on "Inherent soil properties" which was completed by R. Hartmann (Belgium), who also wrote the final version of the chapter and contributed together with G. Monnier (France) to Chapter 4 "Structural state variables", for which K. H. Hartge (Germany) was the main author. J. Bouma (The Netherlands) and L. Hansen (Denmark) combined in the writing of the chapter on "Air and waterflow parameters". "Soil strength and stability" is mainly a contribution of G. Monnier (France); the section on penetration resistance was contributed by K. H. Hartge (Germany) and that on Atterberg limits by D. Gabriels and R. Hartmann (Belgium). The chapter on Soil Morphology was written by J. Bouma and M. Kooistra (The Netherlands). The final version of the book was the result of the editorial work of W. Burke, (Ireland), D. Gabriels (Belgium), and J. Bouma (The Netherlands).

The names of contributors of individual chapters listed above are presented so to allow readers to communicate directly with the authors. Credit for this book is due to the entire group and to every individual who attended the different workshops. A list of names and addresses of workshop participants is therefore included in this book. It was a pleasure and a rewarding task to be chairman of the working group. Herewith I express my gratitude not only to all the scientific workers who contributed so generously to the book, but also to Mr. A. Cole, Dr. G. Prendergast and Dr. E. Culleton of the European Commission, DG VI-Agriculture, who were involved at various stages in the preparation of this book.

M. De Boodt, (Belgium),
Chairman,
Workshop on Soil Structure Methodology.
January 1986.

Names and addresses of Participants

H. J. Altemüller, Institut für Pflanzenernährung und Bodenkunde, (FAL) Braunschweig-Völkenrode, Bundesallee 50, D 3300 Braunschweig, Germany.

G. Antonakopolous, Ministry of Agriculture, Environmental Directorate, 3-5 Ippocratus Street, Athens, Greece.

J. Bouma, Soil Survey Institute (STIBOKA), Starringgebouw, Marijkeweg 11, 6700 AB Wageningen, The Netherlands.

W. Burke, Agricultural Institute, Kinsealy Research Centre, Malahide Road, Dublin 17, Ireland.

G. Chisci, Istituto Sperimentale per lo Studio e la Difesa del Suolo. Piazza d'Azeglio 30, 50121 Firenze, Italy.

A. J. Cole, European Commission, Wetstraat 86, 1040 Brussels, Belgium.

E. Culleton, European Commission, Wetstraat 86, 1040 Brussels, Belgium.

M. De Boodt, Department of Soil Physics, Faculty of Agricultural Sciences, State University of Ghent, Coupure Links 653, 9000 Ghent, Belgium.

P. Dutil, Station de Science du Sol, Route de Montmirail, 5100 Chalons-sur-Marne, France.

D. Gabriels, Department of Soil Physics, Faculty of Agricultural Sciences, State University of Ghent, Coupure Links 653, 9000 Ghent, Belgium.

L. Hansen, Statens Marskforsøg, Siltofvey 2, DK 6280 Højer, Denmark.

K. H. Hartge, Institut für Bodenkunde, Universität Hannover, Herrenhauser Straße 2, D 3000 Hannover 21, Germany.

R. Hartmann, Department of Soil Physics, Faculty of Agricultural Sciences, State University of Ghent, Coupure Links 653, 9000 Ghent, Belgium.

N. Koroxonides, Ministry of Agriculture, Soil Science Institute, Thessaloniki, Greece.

G. Monnier, Station de Science du Sol, Domaine Saint-Paul, B.P. 91, 84140 Montfavet, France.

A. Newman, Rothamsted Experimental Station, Harpenden, Herts, United Kingdom.

G. A. Oosterbaan, Institute for Land and Water Management Research, Marijkeweg 11, Wageningen, The Netherlands.

A. G. Prendergast, European Commission, Wetstraat 86, 1040 Brussels, Belgium.

L. Rixhon, Station de Phytotechnie, Centre de Recherches Agronomiques, Liroux, 5800 Gembloux, Belgium.

A. J. Thomassen, Rothamsted Experimental Station, Harpenden, Herts, United Kingdom.

CHAPTER 1

Site and soil description

The purpose of this description is to provide information which will enable readers to obtain an understanding of the characteristics of sites and soils and of the local climatic conditions, and to compare these characteristics with those of other soils of which they have descriptions or personal knowledge.

In preparing the description one should assume that the readers may have little knowledge of either the soil or its locality. The guidelines should serve for soil physicists and those working in soil structure and related fields.

The description should contain sufficient information for the particular purpose for which the assessment is done and any data on site, soil and climate that is specifically relevant to the type of work. Site description and soil description are treated separately. An adequate description of the site must include details on the more stable features such as location, climate, geological formation, land form, topography, drainage etc., and also information on dynamic characteristics such as crusts, cracks, erosion, runoff. With regard to the soil description it is also necessary to give adequate details of characteristics than can be assessed on site. This includes general descriptions of standard features such as sample types, depth and thickness of horizons, texture, structure, consistency, pores, roots and any other features considered relevant to the type of work.

The description is based on the "Guidelines for Soil Profile Description", prepared by FAO soil surveyors, the greater part of the guidelines being extracted directly from the Soil Survey Manual (USDA, Agricultural Handbook No. 18).

The proposed guidelines to be followed in describing both the site and the soils are set out below.

1.1 Site description

General information:
Location: Indicate region.
Soil classification: Indicate soil names according to the legend FAO-UNESCO Soil Map of the World, Soil Taxonomy, ORSTOM, or CPCS (depending on the region) and local.

Climate:
Climatological data: Indicate precipitation, temperature, relative humidity, windspeed, direction etc.
Weather data: Indicate antecedent and present weather conditions.
Meteorological stations: Indicate stations where data were collected.

Geology:
Parent material: Indicate origin, nature of parent material.
Geological formation: Indicate name of the formation and the geological period.
Mineralogy: Indicate dominant minerals with their relative abundance.

Topography:
Elevation.
Slope: length and gradient
> *Shape:* Concave, convex, linear
> *Pattern:* uniform, irregular, complex.

Geomorphology:
General physiography: Descriptive e.g., plateau, alluvial fan, river terrace, plain ridge, etc.
Landform of the surroundings:

Flat-gently undulating	0—2 % slope
Gently undulating	2—5 % slope
Undulating	5—8 % slope
Rolling	8—16% slope
Hilly	16—30% slope
Mountainous	>30% slope

Erosion and deposition:
> *Water erosion:*
> > *Type::* sheet, rill, gully.
> > *Degree:* Percentage of area affected.
> *Wind erosion:* Percentage of area affected.
> *Deposition:*
> > Overwash (water);
> > Overblown (wind).

Surface characteristics:
> Rock outcrops and stoniness.
> Sealing, crusting, compaction, cracking.

Hydrology and water economy:
> Water table levels.
> Surface runoff or ponding water.
> Drainage:
> > *Internal soil drainage:* slow, medium, rapid.
> > > depth and intensity of gley (rusty iron oxide mottling).
> > *Drainage class:* An integration of runoff, permeability and internal drainage.
> > > *Very poorly drained:* Groundwater table stays near the surface.
> > > *Poorly drained:* The soil remains wet during the greater part of the year.
> > > *Imperfectly drained:* The soil remains wet during a small but critical period of the year.
> > > *Well drained:* The drainage is satisfactory and not too rapid.
> > > *Excessively drained:* The drainage is rapid and no mottling or rust spots occur in the profile.

Flora and fauna:
> *Vegetation:*
> > Forest, pasture, crops, fallow.
> > Composition, Percentage cover, dominant species, rotation system.
> *Soil fauna:*
> > Indicate any evidence of past or present biological activity.
> *Land use:*
> > Describe land use; type of soil tillage; type of water management; type of weed control; type of fertilizers.

1.2 Soil description

Type of sample:
Disturbed or undisturbed.

Thin section sample.
Ring sample.
Other sample—describe.

Horizon:
Depth in cm, upper and lower limit.
Boundary topography.
Specific features: Pan, plough pan, peat, clay layer, gley.
Concretions (including nodules): Indicate quantity, size, type.

Colour:
Munsell notation.
Indicate if dry or moist.

Texture and stoniness:
Texture assessment in the field: Sandy, loamy, silty, clayey.
Indicate size and quantity of stones.

Structure:
Type: Platy, prismatic, columnar, blocky, granular, crumb.

Degree of structural development:
    Indicate if coherent (massive) or non conherent (single grain).
    Weakly, moderately or strongly developed.
Size: Average diameter of aggregates.

Consistency:
The consistency is evaluated according to the moisture content.
Consistency when wet: The soil is at field capacity or higher and the consistency is described in terms
                of stickiness or plasticity.
Consistency when moist: The soil moisture content is between field capacity and air-dry and
                consistency is described as loose, friable, firm.
Consistency when dry: Is described as loose, soft, hard.

Salts:
Descriptive.
Electrical conductivity, pH.

Roots and effective soil depth:
Quantity and size of roots in relation to depth.
Indicate depth to which roots can easily penetrate throughout the year.

1.3 **Literature cited**
FAO Guidelines for Soil Profile Description (1977).
Soil Survey Manual (1951). USDA Agricultural Handbook No. 18, Washington D.C.
Munsell Soil Colour Charts (1954). Munsell Colour Company, Inc., Baltimore, Maryland, USA.
West European Methods for Soil Structure Determination (1967).
ISRIC Soil profile description chart, International Soil Reference and Information Centre,
    Wageningen, Netherlands.

CHAPTER 2

Sampling for soil structure measurement

2.1 Introduction

Sampling for soil-structure measurements may involve removal of samples from a soil profile to be analysed in the laboratory, or placement of in situ equipment at a selected location in the landscape and within the soil. Sampling involves therefore different aspects that cover the selection of: (1) the method to be used; (2) sample dimensions where applicable; (3) sampling locations inside the soil profile; (4) the number of replicates; (5) sampling locations in the field and (6) time of sampling.

Proper consideration of these five aspects will result in better, more representative samples and, therefore, in better data that is intended to define certain soil-structure features. Often, too little attention is paid to these sampling aspects and data are generated without due consideration of method selection and soil conditions. Data, thus obtained, are characterized by a high variability that has only a remote relationship to the soil structure being characterized. The main purpose of using proper sampling techniques is therefore to obtain representative data for a particular soil structural feature, that has a variability which can entirely be attributed to the spatial properties of the feature itself and not to sampling or measurement errors.

Before discussing the five aspects, mentioned above, some attention must be paid to statistical terminology associated with sampling statistics.

2.2 Terminology

The term variability has been used in the introduction, illustrating that different results are often obtained when applying a particular method to a particular soil. This difference can be due to spatial variability of the soil structure feature being measured; but also to many other technical or operational factors. In this context some standard terms are used to express variability (e.g. Petersen and Calvin, 1965):

Bias expresses deviation of the statistical true value from the scientific true value.
Precision expresses variability of an observation around the statistical true value. As the number of samples increases, the precision of the sampling increases. However, high precision does not ensure that bias is absent.
Accuracy is a descriptive term. The term accurate describes a condition that has no bias of any sort and a high precision. The meaning of the terms precise and accurate are graphically illustrated in Figure 1 (Bouma et. al. 1982). Deviation of data from the true scientific value, is referred to as error. Many expressions are presented in the statistical literature for error, such as standard deviations, coefficients of variation, confidence limits, limits of accuracy and others (e.g. Kempthorne and Almaras, 1965; Snedecor and Cochran, 1967; Wilding and Drees, 1982). A further discussion of these statistical terms is beyond the scope of this text. In recent years, geostatistical techniques are increasingly being used. These techniques define the spatial dependence of datapoints, while classical statistics do **not** consider a spatial component (e.g. Journal and Huybregts, 1978). This aspect will be reviewed when considering selection of sampling locations in the field. The five aspects, to be considered when sampling, will now be discussed in more detail.

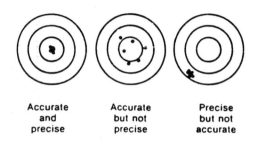

Accurate
and
precise

Accurate
but not
precise

Precise
but not
accurate

Fig. 1 Illustration of the concepts of accurate and precise.

2.3 Selection of method

Once the need has been defined to obtain a measurement of a particular soil-structure feature, a method has to be selected. Soil conditions and operational considerations should play an important role in method selection.

Substantial variability can originate from using the wrong method at the wrong time, at the wrong place or by applying, for instance, a complicated technical procedure, even though only relatively untrained.personnel is available. Some methods use complicated calculation procedures including substantial error even when applied professionally (e.g. Vachaud, 1982). Others yield data directly. A qualitative review of sixteen methods for measurement of the hydraulic conductivity of saturated soil (K_{sat}) and of eleven methods for measurement of K of unsaturated soil (K_{unsat}) was presented by Bouma (1983), emphasizing aspects such as: (1) time needed for preparation, execution and calculations; (2) costs of personnel and materials; (3) complexity and (4) accuracy. Some arbitrarity selected examples will be discussed to illustrate the relevance of method selection:

Example 1: The double-ring infiltrometer is widely and successfully used to measure infiltration rates. The second ring is applied to avoid lateral flow of water from the inner ring. The method works well in sandy soils (e.g. FAO, 1979). However, problems occur in bi-porous clayey soils with cracks where water runs away laterally. As a result, very high infiltration rates are measured in the inner tube. Such rates are much higher than rates occurring in an entirely flooded field. A particular problem occurs when measurements are made in initially dry soil. Short-term soil swelling will not resemble long-term swelling that will occur in a flooded field. A better procedure would be to flood a relatively large area of soil or to excavate a sufficiently large one-dimensional column of soil to be used for an infiltration experiment. In any case, experiments should be made when the soil is naturally wet or has been wetted for some time before the experiment.

Example 2: Sometimes questions have to be answered that cannot be solved by applying existing methods because they are unsuitable. If, for some reason, existing methods are used, questionable results are obtained. In our work, we encountered this problem when asked to measure the vertical K_{sat} of an indurated spodic horizon. Use of sampling cylinders or infiltrometers would have resulted in fracturing of the horizon, making measurement results irrelevant.

It was decided to carefully chip out an in-situ column and encase it in gypsum. Then, the steady infiltration rate in the column was measured while pressure heads were registered simultaneously in soil above and below the spodic horizon (Dekker et. al., 1984). Fortunately, there was time available in this particular project to develop a new method. Often, this time is not available during contract work.

Example 3: This example does not focus on the use of a particular method, but, rather on the occurrence of the phenomenon of "bypass flow" (earlier called: "short-circuiting"). This process describes vertical movement of "free" water along macropores through an unsaturated soil matrix. As such, the process is not discussed in current soil physics text-books. It is important in many soils and it affects results of measurements and adds to the observed variability when methods are used that assume the presence of homogeneous soil. Methods have been developed to measure bypass flow (e.g. Bouma, 1984). Recognition of this phenomenon may help to explain what appear to be erratic, highly variable measurement results at first sight.

Example 4: Different methods are often available to measure the same soil characteristic. In contrast to the first three examples, where some methods produced incorrect results, different methods being considered in this example all produce "good" results. Still, data are different and it is important to define one's purpose in making measurements when selecting one of the methods. Besides, a significant source of variability is created when different methods are concurrently applied. Examples are taken from the work of Dr. P. Stengel (INRA, Avignon, France). He measured soil porosity with two different techniques (Fig. 2). Data in Figure 2A were derived from in-situ measurements determining the volume of a balloon being filled inside a small soil excavation. Data in Figure 2B were derived from large, undisturbed cores of approximately one liter content. The higher values in Figure 2B were due to irregular extensions of the balloon into cavities next to the excavation, which did not occur in the rigid cylinder.

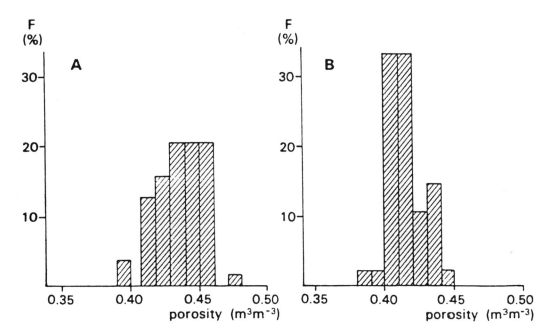

Fig. 2 Comparison of soil porosities obtained with two methods. Method A used the volume of a balloon being filled inside an excavation. Method B used large, undisturbed cores. F is the frequency of measured porosity classes (data courtesy of Dr. P. Stengel, INRA, Avignon, France).

2.4 Sample dimensions

Many measurement procedures use standard sample sizes, because of fixed dimensions of sampling cylinders or of equipment being used. For example, sampling cylinders with a fixed volume of 100 cm³ have been used extensively in different laboratories. Equipment, such as the double-ring infiltrometer or the air-permeameter, comes in standard sizes. There is good justification to vary sample size as a function of soil structure, as a means to reduce variability among replicate measurements (e.g. Bouma, 1983). Soil structure descriptions as made during soil survey can be used to tentatively define representative elementary volumes of samples (REV's), which are the smallest sample-volumes that can represent a given soil horizon by producing an unbiased population of data. To do so, the elementary units of soil structure (ELUS) have to be distinguished (see also chapter 7). These are individual sand grains in sandy soils and natural aggregates ("peds") in aggregated soils. Peds can vary in size up to several liters each in very coarse, prismatic subsoil structures. Even though emphasis in soil structure descriptions is often placed on the solid phase in terms of soil grains and peds, real emphasis should be on the nonsolid phase where transport processes take place. Of course, by describing grains and peds, information is also provided about the pores between them. In addition, pores that do not result from the packing of grains or peds, should be considered separately. Such pores are, for example, root and worm channels with a cylindrical shape. As a general rule we have proposed that REV's should contain at least 20 ELUS but preferably more, or that any sample taken should have a representative number of channels per unit surface area. Defining REV's as a function of field description of soil structure needs to be further investigated. Data by Anderson & Bouma (1973) illustrate the potential of the procedure (Table 1). Unrealistically high K_{sat} values were measured in soil cores in a silt loam containing fewer than twenty ELUS. Values measured with a gypsum-covered column having a volume of 12 liters, averaged 70 cm day^{-1}. The reason for the high K_{sat} values in the small cores is the high and unnatural vertical continuity of cracks between the peds in small samples.

Table 1. Measured hydraulic conductivity of saturated soil (K_{sat}) in soil cores of varying height but with a diameter of 7.5 cm, containing different numbers of elementary units of structure (ELUS). Measurements were made in a silt loam soil with medium sized peds with an average volume of approximately 30 cm^3. The largest sample was a gypsum-covered column of soil with a diameter of 30 cm.

Sample Volume (cm^3)	ELUS (no)	K_{sat} (cm day^{-1})	S (cm day^{-1})
230	8	650	350
330	11	329	320
460	15	100	80
780	26	75	30
12 000	400	70	20

S = Standard deviation

2.5 Sample location in the soil profile
Sampling at regular depth intervals is often applied with good results in relatively homogeneous soils with weakly developed soil horizons. When clear soil horizons exist, however, it is preferable to sample by horizon (e.g. Peterson & Calvin, 1965). A sample containing fragments of two adjacent, and as such quite different, soil horizons, will yield data that are hard to interpret. However, it should be realized that pedological horizons as distinguished in soil survey, are not always good "carriers" of data that are relevant for the particular soil structural interpretation being pursued.   Some pedological distinctions may be irrelevant in this context, while relevant soil-structural aspects may not be reflected in a particular horizon classification. In general, it is advisable to make a soil structure description (see chapters 1 and 7) before taking soil-structure samples for physical analyses. Such samples should preferably be taken in soil layers with a more or less homogeneous soil structure, as observed in the field. Particular attention should be paid to management-induced boundaries such as the lower boundary of the plowlayer, which may occur within a pedological horizon. Another important boundary is the atmosphere-soil interface where crusting and sealing due to rainfall impact may occur in thin layers of only a few centimeters thickness. Selective sampling is desirable here.

2.6 The number of replicates
Having selected the proper method, the sample dimension and the sampling location in the soil profile, the investigator is faced with the question as to how many replicate samples he should take. This question is discussed in detail in any statistical, handbook to which the reader is referred (e.g. Snedecor and Cochran, 1967; Becket and Webster, 1971). The number of samples is a function of the required accuracy: the latter will be higher as the number of samples increase. Graphs have been developed that allow rapid estimation of the number of samples as a function of accuracy obtained (e.g. Wilding and Dress, 1983).

2.7 Sampling locations in the field
So far, the required number of replicates has been discussed, but not their location in the field. Replicate samples can be taken in one soil profile pit by sampling the four profile walls. However, it may be preferable to take individual samples farther apart if an area of land is to be characterized. This does imply, of course, that several soil pits have to be dug. Various sampling schemes for obtaining soil data in the field were recently reviewed by Wilding and Drees (1983). It is advantageous to have a good knowledge of soil and landscape conditions when choosing sampling

sites in an area. Random sampling is suitable only when soil differences are not evident. In other words, when a soil map is available, it is advisable to sample at random within soil mapping units as defined on the soil map. Sampling points may be put in line-transects or grids to allow easier location of sampling points in the field (e.g. De Gruijter and Marsman, 1985).

In classical statistics the distance between replicate sampling points is not important, as long as the individual points are selected at random within the area to be characterized. However, in recent years there have been strong advances in geostatistics, using the theory of regionalized variables (Journel and Hybregts, 1978). These regionalised variables have values that are related in some way to their position. Basically, geostatistical theory states that observations that are located closely together are likely to have a higher probability of resembling one another, than observations that are farther apart. This phenomenon can be mathematically expressed by autocorrelation, by a semi-variogram or by intrinsic random functions (Webster and Burgess, 1980; Bregt, 1985; Nielsen and Bouma, 1985). An idealized semi-variogram is shown in Figure 3. The semi-variance approaches a rather constant value beyond a characteristic distance, which is called the range. Within the range, observations are **not** independent. A certain semi-variance is shown for distance (h) equal to zero. This is called the nugget variance and it represents random sampling error. A known semi-variance can be used to **predict** values at unmeasured locations by interpolation with the kriging technique. A discussion of kriging is beyond the scope of this text. The reader is referred to recent reviews (e.g. Nielsen et. al. 1983; Webster and Burgess, 1983; Nielsen and Bouma, 1985).

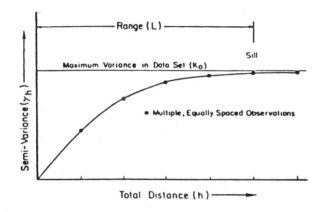

Fig. 3 Idealized semi-variogram for geostatistical studies, illustrating the spatial dependence of separate observation points.

Returning to the topic of this subchapter, which is sampling, attention is focused on the use of geostatistics for optimizing sampling distances. For example, Vieira et. al. (1981) and McBratney and Webster (1983) have showed that use of geostatistics can strongly reduce the number of required replicate samples while still attaining an acceptable degree of accuracy. Application of geostatistical techniques when applying soil-structure measurements is strongly recommended.

2.8 Time of sampling

Many soil structural features change during the various seasons of the year. Water extraction by evapotranspiration in the growing season will, for example, result in cracking of clay soils and in semi-irreversible drying of some peaty and sandy soils. Wetting during fall, winter and early spring will result in swelling of clay soils, Swelling and shrinkage processes are, among other factors, a function of the rate of wetting and of the varying electrolyte content of the soil solution. Usually these long-term processes can not be compressed into a very short period. For example, when measuring the saturated hydraulic conductivity (K_{sat}),\the soil should have been very wet or saturated for several weeks before measurement. Measuring K_{sat} in an initially dry soil is meaningless because short-term rapid swelling will result in different porosity patterns than natural, long-term swelling. In the Netherlands, K_{sat} values or moisture retention curves of clay soils, are only measured on samples that have been taken in the period February to March in early spring when soils have been naturally wet for several months.

Determinations of soil structure stability are also very sensitive to the time of sampling, if only because of the changing microbiological activity during the year.

The above examples are just meant to illustrate the problem being considered here. More examples could be given. In general it is advisable to always consider the aim of any test to be made and to establish the seasonal range of the particular property being considered. The time of sampling should correspond with the time of the year in which the particular property being considered, is thought to be most important.

2.9 Literature cited

Anderson, J. L. and J. Bouma, 1973. Relationships between hydraulic conductivity and morphometric data of an argillic horizon. Soil Sci. Soc. Amer. Proc. 37: 408-413.

Beckett, P. H. T. and R. Webster, 1971. Soil variability; a review. Soils and Fertilizers. 34: 1-15.

Bouma, J. 1983. Use of soil survey data to select measurement techniques for hydraulic conductivity. Agric. Water Managem. 6 (2/3): 177-190.

Bouma, J. 1984. Using soil morphology to develop measurement methods and simulation techniques for water movement in heavy clay soils. *In:* J. Bouma and P. A. C. Raats (eds.). Water and Solute Movement in Heavy clay soils. Proceedings of an ISSS Symposium. ILRI, Publ. 37, Wageningen, Neth. 363 p.

Bouma, J., R. F. Paetzhold and R. B. Grossman, 1982. Measuring hydraulic conductivity for use in soil survey. Soil Conserv. Serv. USDA. Soil Survey Investigations Report no. 38, 14 pp.

Bregt, A. K. 1985. Geostatistical techniques and spatial variability of soil physical properties. *In:* T. Woodhead (Ed.) Proc. of Int. Workshop on Physical Aspects of Soil Management in rice-based cropping systems. IRRI, los Banos, Philippines (in press).

Dekker, L. W., J. H. M. Wösten and J. Bouma, 1984. Characterizing the soil moisture regime of a Typic Haplohumod. Geoderma 34: 37-42.

FAO, 1979. Soil survey investigations for irrigation. FAO Soils Bulletin 42. Rome, Italy.

De Gruijter, J. J. and Marsman, B. A. Transect sampling for reliable information on mapping units. *In:* Nielsen, D. R. and Bouma, J. (Eds.), Soil spatial variability. Proceedings of a Workship of the ISSS and the SSSA, Las Vegas, USA, 30 Nov.—1 Dec. 1984. p. 150—166. Pudoc, Wageningen.

Journel, A. G. and Ch. J. Huybregts. 1978. Mining geostatistics. Academic Press. London.

Kempthorne, O. and R. R. Allmaras, 1965. Errors of observations. Chapter 1 in: Methods of Soil Analysis. Part 1. C. A. Black (ed.) Am. Soc. of Agronomy. Agronomy Series no. 9: 1-23.

Mc Bratney, A. B., and R. Webster. 1983. How many observations are needed for regional estimation of soil properties? Soil Science 135: 177-183.

Nielsen, D. R. and J. Bouma (Eds.) 1985. Soil spatial Variability. Proc. of an ISSS-SSSA workshop, Las Vegas USA. Pudoc Wageningen, the Netherlands.

Petersen, R. G. and L. D. Calvin, 1965. Sampling. Chapter 5 *In:* Methods of Soil Analysis. part 1. C. A. Black (ed.) Am. Soc. of Agronomy. Agronomy Series no. 9: 54-73.

Snedecor, G. W. and W. G. Cochran. 1967. Statistical Methods (6th Ed.) Iowa State Univ. Press. Ames. Iowa. 593 pp.

Vachaud, G., 1982. Soil physics research and water management. Whither Soil Research. Proc. ISSS Congress, India: 32-59.

Vieira, S. R., D. R. Nielsen and J. W. Bigger, 1981. Spatial variability of field measured infiltration rate. Soil Sci. Soc. Am. J. 45: 1040-1048.

Webster, R and T. M. Burgess. 1980. Optimal interpolation and isarithmic mapping of soil properties. III Changing drift and universal kriging. J. Soil Sci. 31: 505-524.

Webster, R. and T. M. Burgess, 1983. Spatial variation in soil and the role of kriging. Agr. Water Managem. 6: 111-123.

Wilding, L. P. and L. R. Dress, 1983. Spatial variability and pedology. *In:* Pedogenesis and Soil Taxonomy. L. P. Wilding, N. E. Schmeck and G. F. Hall (eds.): Developments in Soil Science IIA-Elsevier, Amsterdam: 83-113.

CHAPTER 3

Inherent soil properties

Inherent soil properties are understood to be those properties of the solid or mineral constituents of soil which scarcely change with time, but which strongly influence the physical and structural behaviour of the soils.

The other components or constituents of the soil, namely the water solution and soil air, are largely influenced by:
—changes in the structural status of the soil
—external factors such as rainfall, evaporation, capillary rise of water, drainage, crop growth etc. . . .

Knowledge of the composition of the solid phase is necessary before any other soil physical property is determined or structural status evaluated.

It is proposed that the following inherent soil properties should be determined:
— texture, or particle size distribution of the mineral matter of the soil
— particle density of the mineral soils
— clay mineralogy of the particles < 2 μm
— chemical components that influence physical condition.

3.1 Particle size distribution

3.1.1 Introduction
Soil particles comprise the solid phase of the soil and can differ from one another in composition and size. These particles may be organic or inorganic, crystalline or amorphous and they largely determine the behaviour of the soil.

The method described hereafter applies to the inorganic particles because in most cases they largely represent the solid phase.

The particle size distribution of a soil is one of the most stable soil characteristics in ordinary time and under normal conditions and expresses the proportions of the various sizes of particles which the soil contains.

The procedure to separate and measure different fractions is called the particle size analysis or mechanical analysis. The result of this analysis is often termed texture and refers only to particles smaller than 2 mm in diameter. Material larger than 2 mm should also be quantified.

There is not yet a universally accepted scheme for classification of particle size and the various criteria used by different workers are rather arbitrary.

For soil classification, it is recommended that three main fractions be used, sand (50—2000 μm), silt (2—50 μm) and clay (< 2 μm), thus enabling the composition to be represented on a triangular system.

As in different laboratories the size limits of different fractions in use are sometimes quite different it is recommended that a graphical representation of the particle size distribution which is known as the summation curve, be given.

Estimation of particle size by sedimentation is considered to be more accurate using the pipette method than the hydrometer method, so the pipette method is recommended as the reference method.

3.1.2 Pipette sampling

Principle

The method described here is limited to sieving and sedimentation procedure. The latter consists of removing with a pipette a sample of known volume from a given depth in a water suspension of dispersed soil at a specific time after sedimentation has begun.

Equipment

Pipette sampling apparatus, 1 liter graduated cylinder, large temperature controlled water bath, balance, oven, set of sieves (2000 µm and 50 µm) with cover and pan, sieve shaker with timer, evaporating dishes, beaker (250 ml), heating system (hot plate or water bath), stopclock, desiccator, plunger.

Chemical products such as hydrogen peroxide (H_2O_2), hydrochloric acid (HCl) and sodium metaphosphate ($NaPO_3$).

Procedure

Dry and crumble the field sample. Separate the fine earth fraction by sieving through a 2 mm sieve.

For effective fractionation the soil (<2 mm) must be dispersed in water. This may require dissolution of free carbonate by acidification (HCl), of organic matter by oxidation (H_2O_2) and of sparingly soluble salts by washing.

Afterwards the particles greater than 50 µm are removed by wet sieving. The fraction less than 2 µm is determined by sedimenting the suspension of dispersed particles <50 µm for a specified time and determining the particle concentration at a specified depth by volumetric sampling.

Calculation

Determine:
— the oven-dry weight of the sand separate (50—2000 µm)
— the oven-dry weight of the clay fraction (<2 µm) calculated from the amount taken with the pipette with volume V and corrected for the weight of dispersing agent
— the oven-dry weight of the silt fraction (2—50 µm) by difference.

Each fraction can be expressed as percentage of the oven-dry mineral portion of the soil. As other size limits are sometimes used, always summarise the particle size distribution data on a summation curve. In such a curve the abscissa represents the particle size on a logarithmic scale and the ordinate the percentage of materials smaller than any particular size.

Time required

It takes several days before the particle size distribution of a soil sample is obtained.

Cost

The purchase of the equipment is expensive. Running costs are low.

Accuracy

Good reproductibility requires careful standardisation of procedure. In using Stoke's law it is assumed that the particles are spherical which is not in accord with reality. Moreover, soil particles are not all of the same density. For these reasons the analysis yields only approximate results.

Remarks

— Oxidation of organic matter with H_2O_2 in the presence of calcium carbonate is inefficient. It is preferable therefore to decalcify before peroxidizing.
— The method is arbitrary since it attempts to divide a continuous array of particle sizes into distinct preconceived fractions.
— Specific size fractions may be needed for certain purposes. However, it is recommended that sizes of 2 mm, 50 µm and 2 µm should always be included.
— In calcareous soils, dissolving the $CaCO_3$ leads to inaccurate results. In such soils the particle size distribution is determined without destruction of the calcium carbonates and the dispersion is obtained by substituting the exchangeable Ca^{++}-ions by K^+-ions, followed by shaking in the presence of sodium hexametaphosphate.

3.1.3 Hydrometer

The hydrometer is a widespread alternative method for measuring the amount of material in suspension without having to take pipette samples, evaporate them to dryness and weigh.

The density of the suspension measured at given depth as a function of time, is used to calculate the concentration of the soil in the suspension.

The greatest disadvantage, among others, of this method with respect to the pipette method lies in the fact that the concentration is influenced by the density of the soil particles which can vary between soils.

References

Day, P. R. 1965. Particle fractionation and particle size analysis. *In:* Methods of Soil Analysis, I: (ed. C. A. Black): 545-567. Agronomy no 9, Am. Soc. Agron., Madison, Wisconsin.

Dewis, J. and Freitas, F., 1970. Physical and chemical methods of soil and water analysis. FAO Soils Bulletin, 10: 1-275.

U.S. Department of Agriculture, 1972. Soil survey laboratory methods and procedures for collecting soil samples. Soil Survey Investigations, Report No 1, 1-63.

West-European Methods for Soil Structure Determination, 1967. Laboratory information on soil components. Grain size distribution, IV, 11-IV. 91., Gent, Belgium.

3.2 Particle density

3.2.1 Introduction

The particle density of soils refers to the density of the solid particles. It is defined as the ratio of the total mass of the solid particles to their total volume excluding pore spaces between particles.

Although considerable range may be observed in the density of the individual soil minerals the values for most mineral soils vary between narrow limits viz. 2.60 and 2.75.

Surface soils usually have lower particle density than subsoils due to the presence of organic matter which weighs much less than an equal volume of mineral solids.

3.2.2 Submersion method

Principle

Oven dried soil material is weighed in air and submerged in xylene $C_6H_4(CH_3)_2$.

Procedure

Approximately 20-40 g of moist soil is passed through a 2 mm sieve to form small wormlike aggregates. These are put into a weighing dish, dried at 105°C and weighed to 0.1% accuracy. Then the dish is filled up with xylene so that all aggregates are covered, stirred cautiously with a glass rod, placed into a vacuum desiccator and evacuated for 5 minutes to expel the rest of the air.

The dish is then put in a wire basket and weighed submerged in a xylene bath of at least 1 litre content. Xylene level in this vessel is kept constant and the vessel must not be touched with the hands to avoid temperature change.

The weighing dish is emptied now and its submerged weight determined.

At the end of each series a piece of solid (quartz, brass) is weighed in the same way in order to avoid errors caused by changes in xylene density due to impurities or temperature change.

Calculation

Particle density (dp) is given by:

$$dp = \frac{d_1 (W_s - W_v)}{(W_s - W_v) - (W_{sl} - W_{vl})} \quad (kg.m^{-3})$$

where:
d_1 = density of submersion liquid ($kg.m^{-3}$)
W_s = weight of oven-dry sample with weighing dish (kg)
W_v = weight of weighing dish (kg)
W_{sl} = weight of sample and dish submerged (kg)
W_{vl} = weight of dish alone submerged (kg)

d_l should be controlled by submerged weighing of a piece of quartz or brass or similar material. The submerged weight of the latter in water having been previously determined.

$$d_l = \frac{d_w \cdot (W_b - W_{bl})}{(W_b - W_{bw})}$$

where: d_W = density of water at working temperature (from tables)
W_b = weight of control material in air
W_{bl} = weight of same material submerged in liquid
W_{bw} = weight of same material submerged in water

Equipment
— balance which allows a wire basket to be hung by a single thin wire so that the weighing dish is submerged, accuracy 0.1% of sample weight
— weighing dishes
— vacuum desiccator and evacuation pump
— drying oven (105°C)
— Xylene

Time required
Working time for one sample requires 4 weighings and some manipulation such as drying, evacuating and later on submerging the sample. All, including calculation, ± 10 min. if a modern quickworking balance is used and the working routine is well established. Drying and evacuating time not included.

Cost
Cheap.

Accuracy
Standard deviation with well mixed samples, split up for parallel measurement is 5 kg.m^{-3}.
    Standard deviation of unmixed samples might be ten times greater.

Advantages/Disadvantages
— With the submersion method it is easier to obtain low scatter of values than with pycnometers, because volume determination is replaced by the more accurate weighing.
— Because of the large xylene volume, temperature induced errors are minimized.
— Xylene wets soil particles easily and allows them to sediment more quickly than does water.
— Disadvantage is smell of xylene.

References
Blake and Hartge, 1984. *In:* Asa-Method book (in preparation).
East European Method Book ISSS, 1968. Komm, I., VEB-Deutscher Landwirtschaftsverlag Berlin.
West European Methods for Structure Determination, 1967. Determination of particle density, V, Gent, Belgium.

3.3 Clay mineralogy

3.3.1 Introduction
The fraction $> 2\mu m$ in soils generally does not contain colloidal material and from the point of view of soil structure its mineralogical composition is of minor importance, although particle shape (angular or rounded) may be a factor in packing primary particles and hence in bulk density.

Mineralogy of the fraction $< 2\,\mu m$ however influences how much water is taken up and the volume of the colloid fraction depends in part on the swelling mineral content. The mineralogy of clay, as well as the amount of it, is therefore significant in soil structure study.

X-ray diffraction is the most informative single technique in clay mineralogy.

3.3.2 X-ray diffraction

Principle

It is based on the diffraction of X-rays from the crystalline mineral components. Since no two minerals have exactly the same interatomic distances in three dimensions, the angles at which diffraction occurs will be distinctive for a particular mineral.

Equipment

Commercial X-ray diffraction equipment.

Procedure

1. **Preparation of the sample**
— removal of flocculating and cementing agents such as soluble salts, calcium carbonates, organic matter and free iron and aluminium oxides
— separation of the clay fraction ($<2\,\mu m$). It is isolated by dispersion and sedimentation.
— following clay samples should be prepared:
    a. saturation of the exchange complex with $MgCl_2$ to obtain a Mg-saturated air-dried sample: for general information.
    b. solvation of the Mg-saturated clay with ethylene glycol (ethanediol $CH_2(OH).CH_2OH$) to distinguish between swelling clay minerals and vermiculite.
    c. saturation of the exchange complex with KCl to obtain a K-saturated air-dried sample: for general information.
    d. heating the K-saturated sample to 500°C to determine kaolinite and chlorites.

2. **X-ray examination of the samples**

Calculation

The method allows a qualitative interpretation by either:
— direct comparison of diffraction patterns of unknown samples with patterns obtained from known minerals.
— measurement of diffraction spacings and comparison of these spacings with known spacings of standard minerals.

Time required

Preparation of the sample takes a long time in comparison with the effective X-ray diffraction technique.

Cost

Equipment is expensive.

Accuracy

Very good.

Advantages

The only accurate method. No chemical analysis is necessary.

Disadvantages

Only a qualitative analysis or, at best, semi-quantitative. The removal treatments may alter the properties of some clay minerals in the sample.

Remarks
In comparison with the particle size analysis extra care must be taken in dispersion to preserve the clay minerals present. As an example use sodium acetate instead of hydrochloric acid to remove calcium carbonate.

References
Brown, G., 1961. The X-ray identification and crystal structures of clay minerals. Mineralogical Society London, 1-594.
Cailliere, S. and Hénin, S., 1963. Minéralogie des argiles. Masson et Cie, 1-294.
Joint Committee on Powder Diffraction Standards, 1975. Powder diffraction file. Search manual. Hanawalt method. Inorganic Compounds. Publication SMH-25, 1-994.
Soil Science Society of America, 1977. Minerals in soil environments. Ed. R. C. Dinauer, 1-948.
Whiting, L. D., 1965. X-ray diffraction techniques for mineral identification and mineralogical composition. *In:* Methods of Soil Analysis, I: (Ed. C. A. Black): 671-698. Agronomy no 9, Am. Soc. Agron., Madison, Wisconsin.

3.4 Chemical components influencing soil structure

Although the reasons for the influence are not always understood, it has been generally accepted that soil structure and its response to stresses are influenced by calcium carbonate, gypsum, organic matter, free sesquioxides (iron and aluminium), salt content and exchangeable bases especially sodium.

It is therefore anticipated that in many studies of soil structure estimates will be made of some of the above components in the soil.

It is known that the results of such analyses may be influenced by the method used, so the reference methods described in EUR 6368 "Workshop on standardization of analytical methods for manure, soils, plants and water" from the Commission of the European Communities should be followed.

A. **No further listing is needed here for:**
— pH H_2O and pH KCl
— conductivity of saturation extract
— total $CaCO_3$
— organic matter
— cation exchange capacity
— exchangeable cations: Ca, Mg, K, Na.

B. **Determination of free iron and aluminium oxides**
Since the nature and the amount of the extracted compounds depend on the extraction method used, the following reference method is proposed: "Dithionite—citrate—bicarbonate extraction".

References
US Department of Agriculture, 1972. Soil survey laboratory methods and procedures for collecting soil samples. Soil Survey Investigations, Report No. 1, 1—63.
Workshop on standardization of analytical methods for manure, soils, plants and water. Editor: Professor Dr. A. Cottenie, EUR 6368 EN, Commission of the European Communities.

C. **Determination of gypsum**

(a) **Low gypsum content**
The method is based on dissolving gypsum in water and on the gravimetric determination of SO_4 in the extract.

Reference
US Department of Agriculture, 1972. Soil Survey laboratory methods and procedures for collecting soil samples. Soil Survey Investigation Report No. 1, 1-63.

(b) **High gypsum content**

The soil is treated with a H-saturated strong acidic (sulfonic) resin. All cations are exchanged by H of the resin. The anions are transformed in the corresponding acid. The total amount of titrated acid is corrected for the amount of Cl and gives the amount of SO_4.

Reference

De Conick, F. (in press). Handbook for chemical soil analysis. Laboratory for Soil Science. Rijksuniversiteit Gent, Belgium.

CHAPTER 4

Structural state parameters

4.1 Introduction

If the physical condition of a soil is to be determined, investigations usually begin with parameters characterising the structural state. The term "state" in this context means properties, which (1) seem to be static in the sense that they do not change rapidly with time, and which (2) do not describe processes nor do they influence processes directly.

This concept seems to be classical and well established in soil science, although from the point of view of physical properties it is questionable. Initially one tends to include in this group all those properties that relate to soil geometry, but further consideration shows that such classification is not strictly correct.

The main reason for this is that all geometric conditions are subject to environmental influences. This is relatively clear for properties such as porosity or bulk density, but in principle it is also valid for aggregate size, and even for grain size. All of these parameters are affected by the amount of load that is or was applied, and therefore they include aspects of stability, which obviously are not static in the present context.

On the other hand there is a close relationship between soil geometry and properties that are considered functional, such as permeability to water or air. Even these properties could be considered to be states of the soil-matrix, because if the geometry of the matrix is changed—for example by compression—these properties are also changed. Thus they are essentially properties of the internal geometry of the soil and therefore properties of state.

Both examples may serve to explain the difficulties encountered when trying to define, what are the state properties of soil. It is clear also that these properties that comprise the state of soil structure must not be confused with "state parameter" as defined in thermodynamics.

When structural state is in question, there are specific properties that can only be measured on samples, and others that can be measured equally well on the soil in-situ. Here also it is difficult to give a strict definition.

Thus when deciding on whether to use field or laboratory methods it is necessary to take into account initially the type of apparatus available and the number of replicate determinations that are possible. Finally there is a question of economy. Where a choice is possible the following facts should be considered:

In most cases deviations from a mean value decrease, if sample size increases. This encourages the use of field methods, if they imply that the sample size is actually bigger than that used in laboratory measurements. Smaller samples are recommended, if changes within small distances are to be investigated.

It may, however, be less laborious to take a large number of small samples evenly distributed on a given area, than to seek the same degree of representativity by in-situ measurements on bigger samples. Furthermore it should be kept in mind that the extraction of small samples tends to bias the measurement more than the extraction of big ones by changes of state created within the sample.

This occurs with core samples as well as with excavated ones. Finally some determinations such as grain size distribution and particle density can only be carried out in the laboratory.

Ultimately it is essential to recognise that there are two different problems that may influence the

23

result: sampling and measuring. The term "field method" implies that the measurement is performed in the field; sampling is not avoided in this case but is implicit in the choice of the measuring position and the number of replicated measurements distributed on the area which is to be characterized.

Finally it emerges that the first point to be considered when measuring state properties, should be the decision whether "field" or "laboratory" methods are to be used.

The decision should be taken very carefully and explained as precisely as possible, because a result is only as good as the choice of sampling method and the processing of the sample.

In the previous chapter, the measurement of more permanent characteristics was discussed. This chapter deals with properties of the structural state that are more liable to change. Included are bulk density and porosity, aggregate size distribution, soil water content and water retention properties.

4.2 Bulk density and porosity

4.2.1 Stone-free soils
Principle:
Volume of sample is known and constant, only the dry mass has to be determined. Porosity is calculated from bulk density and solid density.

Procedure:
Core samplers (made from stainless steel tubes e.g., i. d. = 88 mm, h = 35 mm) sharpened at one end are driven into soil with as few blows as possible. Hammering should be done with a hammer on a sliding rod.

Cores are dug out, cut smooth at both ends, cleaned from outside and weighed. Aliquot of contents is dried at 105°C and water content calculated.

The bulk density (d_B) and porosity (n) are calculated.

Equipment:
1. Core samplers with tightly fitting covers.
2. Sampling device.
3. Hammering cap (or anvil with sliding rod)
4. Sliding hammer 5 kg.
5. Spade.
6. Knife (straight blade, length 1.5 times diameter of core sample).
7. Balance weighing 0 to 500 g, 1% accuracy.
8. Weighing dishes for water content determination.
9. Drying oven (105°C).

Calculation:

$$d_B = \frac{\text{dry mass (m)}}{\text{vol}}$$

$$m = \frac{\text{mass of wet soil (core)} . \text{mass dry soil (aliquot)}}{\text{mass wet soil (aliquot)}}$$

$$\text{vol} = \text{core volume}$$

$$n = 1 - \frac{d_B}{d_P}$$

d_P (particle density) can be obtained from tables (quartz = 2.65) or determined separately by submerged weighing of dried sample in xylol, or by using pycnometer-technique.

Results for n are given as a percentage (e.g. 45%) or as volume fraction (e.g. 0.45).

Time:
Sampling: 5 to 7 minutes per core from soil surface.

Laboratory procedures:
Including all weighings dependent on type of balance ca. 5 minutes.

Calculations:
5 minutes.

Cost:
Relatively cheap.

Accuracy:
Standard deviation for $n \pm 3\%$ or 0.03 with 200 cm^3 cores — for $d_B \pm 0.08$ (g.cm^{-3}).

Advantages/Disadvantages:
Easy and correct volume determination. Not applicable in stony soil. Difficult in very loose, freshly tilled soil and in sands. Samples and hammers must be carried to sampling points.

Complimentary remarks:
Larger cores tend to reduce standard deviation, shorter wider cores (d:h >1) minimize sampling effects close to core walls; wide cores (d:h > 2.5) increase risk of sample loss in soils having low densities.

Hammering should be done with as few blows as possible and with a heavy sliding hammer (5 kg) to reduce vibration.

Core samplers should be given a long sharpened shoulder for easy penetration into soil.

4.2.2 Excavation in stony soils

Principle:
Sample volume and dry mass must be determined separately. Volume is determined by measuring the excavation made to get the mass. Bulk density and porosity are calculated as before.

Procedure:
The soil surface is smoothened and the sample is excavated. Volume may be obtained by filling the excavation with sand or by direct measurement. All excavated soil material is collected in a bag or pail to be weighed. After weighing, the material is thoroughly mixed, an aliquot taken and stored carefully for water content determination. The rest is discarded.

The excavation is filled with medium-size pure sand to give a smooth surface as before sampling. Sand is added from a calibrated cylinder, so that volume can be read directly from the cylinder.

If sand is not used, the distance (height h) of the smoothened surface is measured from a reference level before excavation and the distance to the bottom of the excavation after sampling. For this measurement a grid is used so that every measurement of depth can be related to a unit of horizontal area.

Equipment:
 1. Plastic bags.
 2. Spring balance.
 3. Spade.
 4. Several spoons.
 5. Pail of pure medium sand (5l).
 6. Weighing dishes, laboratory balance.
 7. Oven (105°C).
 8. Calibrated cylinder (500 cm^3) to measure sand volume.
 9. Grid for distance-measuring.
10. Three blocks to position grid.
11. Ruler to measure depth of excavation.

Calculation:
Bulk density d_B is calculated from dry mass (m) and its volume (vol).

$$m = \frac{\text{mass of wet soil (total) . mass of dry soil (aliquot)}}{\text{mass of wet soil (aliquot)}}$$

vol = vol. of sand from measuring vessel that has been used to fill the excavation or sum of distances (h) before excavation substracted from sum of distances after excavation, times unit area (A):

$$\text{vol} = \Sigma \, \text{h(after excavation)} - \Sigma \, \text{h (before excavation) . A}$$

Time:
Excavation and volume determination with sand — 15 minutes. Excavation and measurement of distances from reference level — 30 minutes.

Calculations:
10 minutes.

Cost:
Relatively cheap.

Accuracy:
Standard deviation for sand-fill-method ± 5% for n (vol. approximately 900 cm³); ± 0.13 for d_B.
  Standard deviation for height measurements ±7% for n (vol. approximately 900 cm²); 0.2 for d_B.
  Accuracy may be increased by increasing excavation volume.

Advantages/disadvantages:
Both methods applicable in all soils.
  Both methods applicable if samplers and driving devices are not available. Less equipment to carry than with samplers especially when height measurements are used.
  Less accurate and more time consuming than core sampler method.

Complimentary remarks:
1. Since 85% sand may be recovered, 5 l sand is sufficient for 7 determinations of excavations > 2000 cm³.
2. For measuring heights a ruler should be used that ensures vertical measurement.
3. Excavations should be made only where the surface is smooth. Width through-out sampling depth should be as uniform as possible, especially when surface samples are taken.
4. Methods of coating clods with paraffin, wax or resin are not recommended because larger pores at the surface will not be included and the sample will tend to expand before coating because of relaxation of horizontal stress.

References:
West European Methods for structure determination Ghent, (1967).
East European method-book ISSS, Komm. I, VEB-Deutscher landwirtschaftsverlag Berlin, (1968).
Methods of soil analysis, American Soil Agronomy, Madison, Wisc. (1965).

4.2.3 Gamma radiation techniques
The interaction of gamma radiation with soil provides a method for measuring the total density of soil. By suitable calibration, measurements of either transmission or scattering of gamma radiation can be used to estimate bulk density.

The most commonly used instruments (depth and surface probes) are back-scattering devices. To describe the physical state of the upper layer (0—40 cm) of the soil and its evolution under the influence of hydrological and mechanical stresses the two-well gamma probe, which is an absorption transmission device, is a very promising technique.

If the soil matrix remains invariant with the exception of its soil water content, gamma radiation can also be used to measure soil water content (see 4.4.4).

4.2.3.1. Back-scattering technique

Principle:
The technique employs a single probe containing both gamma source, mainly ^{317}Cs, and detector — either GM tubes or a scintillation crystal — separated by shielding in the probe. Changes in bulk density produce changes in both the number of gamma rays scattered towards the detector and the proportion absorbed before they reach the detector. The combined effect of scattering and absorption is that a smaller number of gamma rays will reach the detector as the surrounding material becomes more dense.

Equipment:
This technique can be used either at the soil surface (surface density gauge) or placed in a hole (depth density gauge) depending on the design of the equipment.

Equipment consists of a probe, containing the gamma radiation source and detector, shield and ratemeter or scaler and timer.

Soil sample containers to remove samples for soil water content determination.

When using the depth density gauge, access tubes and a device for boring holes in the soil are necessary.

Procedure:
1. Surface density gauge.
    Place the gauge on a smooth soil surface and ensure an intimate contact. Measure the count-rate and the standard count-rate according to the manufacturer's instructions.
2. Depth density gauge.
    Install an access tube to the desired depth. Lower the probe to the desired depth and measure the count-rate. Determine the standard count-rate in its carrying shield.

Calculation:
The count-ratio is converted to wet bulk density by means of a calibration curve. To calculate the dry bulk density the water content of the soil directly below the gauge or of the soil surrounding the site of the measurement has to be determined.

Time required:
The measuring time is very short and the bulk density can easily be calculated. For practical use the instruments are empirically calibrated in the laboratory or in the field and this takes quite some time.

Cost:
Very high. Installation and measurements require a well trained operator.

Accuracy:
Depends largely on the calibration curve.

Advantages:
Minimum disturbance of the soil. Short time required for sampling. Accessibility to subsoil measurement with minimum excavation. Possibility of a continuous repeated measurement at the same point (depth).

Disadvantages:
A spherical sample, which varies from about 20 to 75 cm in diameter depending on characteristics of the sampler and the soil, is tested.

The scattering technique requires a greater source strength than the transmission technique. There is a high radiation hazard.

4.2.3.2. Transmission technique

Principle:
Two small access tubes are placed in the soil so that the distance separating them is fixed and well defined. A gamma-radiation source (^{137}Cs) is placed in one tube and a detector is lowered to the same depth in the other tube.

The absorption of emitted photons is an exponential function for a given soil of its volumetric water content and of the dry density of the medium traversed.

Equipment:
A gamma transmission double probe with scaler and timer. Drilling equipment and tubes for the bore holes. Soil sample containers to collect samples for soil water content determinations.

Procedure:
The source and the detector are placed at the same depth in the soils in previously drilled holes 10 to 15 cm apart (1 mCi) or 20 to 30 cm apart (5 mCi).

The moisture content is measured by weighing samples taken while drilling the holes.

The number of impulses received by the counter is transformed into bulk density by means of a calibration curve established for each scanning interval.

Calculations:
Once the calibration line is established, the calculations for keeping account of the variation in water content of the layer being examined are very simple.

Time:
The drilling of the prepared holes is fairly time consuming, but the scanning time is short (1 minute per depth scanned). The acquisition of the calibration curves takes quite some time.

Cost:
The acquisition of the density probe and the creation of a calibration bench are fairly expensive.

Accuracy:
The sources of error arise from:
The uncertainty inherent in the random character of photon emission. In normal conditions counting time should be of the order of one minute.
Errors of temperature (very small).
The quality of calibration.
Lack of paralellism of the previously drilled holes (error of scanned thickness).
Interaction between layers having very different densities while scanning in the vicinity of the interface.

Advantages:
Measurements are possible irrespective of the physical state (degree of compaction or humidity). Good resolution with depth allows description of very different porosities in a profile.
The length of sampling (30 cm) is sufficient to overcome several sources of heterogeneity in the soil. Precise characterization of overall effect of tilling techniques and phenomena of consolidation as the transmission technique has the advantage of confining the sample to a soil horizon of a few centimeters in the vertical dimension between the probes.

Disadvantages:
The necessity to measure moisture by weight.
Incompatible with stony soils.

Difficulties result from lack of parallelism of the predrilled holes for deep measurements and (or) in dry and fissured soil.

Remark:
The same device can be used to measure the distribution of the soil moisture content in the top layer with the aid of a calibration chart. Devices based upon the same principles of scattering and attenuation have been developed and proved to be useful for laboratory investigations. It is frequently used to monitor accurately and at regular times the soil water content in soil attenuation, whose density does not change with water content (see 4.4.4. Gamma-ray)

References:
Blake, G. R., (1965). Bulk density. *In:* "Methods of Soil Analysis" (Ed. C. A. Black). Agronomy 9, Am. Soc. Agron., Madison, Wisconsin.

IAEA. (1976). Tracer Manual on Crops and Soils. Technical Reports, Series No. 171.

INRA. (1984). Double Sonde Gammamétrique L.P.C.—INRA. Science du Sol, Centre de Recherches Agronomiques d'Avignon.

4.2.4 Bulk density of aggregates

Objective
To evaluate the porosity of a material at one level of organisation where it results principally from the assembly of the elementary constituents (texture) or the moisture (state of swelling) content.

Principle:
The bulk density is determined on small fragments taken at some moisture content and of which the minimum size depends on the granulometric composition of the material. They are impregnated with kerosene and their volume is measured by hydrostatic upthrust in the same liquid.

Equipment:
A balance graduated in milligrams.

Procedure:
The aggregates sieved to the selected size are submerged in kerosene for 24 hours. They are then quickly dried on blotting paper just until the peripheral film of kerosene is eliminated. They are then placed in a container of known volume and weight and their volume is measured by hydrostatic upthrust in the kerosene.

Time—Cost:
About 40 basic results per day.

Accuracy:
The mean standard variation of the bulk density is less than 0.02 which corresponds to less than one point of porosity.

Advantages:
The method is rapid and precise, not very difficult and does not require special materials.

Disadvantages:
In the case of sandy texture there is a danger of losing the kerosene contained in the largest pores at the time of drying. The size of aggregates taken for the measurement should be increased to limit the error.

Manipulation is difficult while bringing clayey aggregates to a moisture content corresponding approximately to the swelling potential.

Areas of utilization:
Establishment of shrinkage curves.
Analysis of porosity.
Analysis concerning the phenomena of compaction.

Variations:
Measuring density of clods by coating or by paraffin.

References:

Monnier, G., Stengel, P., Fies, J. C., (1973). Une méthode de mesure de la densité de petits agglomérats terreux; application à l'analyse des systèmes de porosité du sol. Ann. Agon., 24(5), 533—545.

Fies, J. C., Stengel, P., (1982). La densité texturale des sols. I-Méthode de mesure. Agronomie, 1(8), 651—658.

4.3 Aggregate size distribution

4.3.1. Dry sieving

Principle:
Air dry soil is sieved to separate aggregates of different sizes. The weight (or percentage) of aggregates in the several size classes are compared.

Procedure:
Soil samples (2—3 kg) are taken (with a spade or cylinder) at the desired depth and carried in wooden or plastic frames or stacks (not bags) and transported with minimum agitation and without compression. The sample is carefully broken to form clods with a maximum size easily manageable for the intended analysis and air dried. The air dried sample is placed on the top of a set of sieves; not more than 100 g of soil per 100 cm^2 of sieve surface should be used. The set of sieves is agitated or rotated for several minutes or repeatedly hit with a rubber hammer. Each soil fraction on each sieve is weighed.

Equipment:
Set of sieves with mesh sizes between 8 mm and 2 mm for clay and loam soils and between 8 mm and 1 mm for sandy soils; sieving machine; rubber hammer.

Calculations:
Evaluation of the size distribution is done by drawing a summation curve from which the mean weight diameter (MWD) can be calculated.

Remarks:
The choice of sieves and of physical sieving work is almost unlimited. The sample must not be sieved before it is air-dry. Differences arise from unequal sieving action. As clod or aggregate size distribution varies with time, especially on freshly tilled plots, samples should frequently be taken at moisture conditions when the stability is at a minimum. Therefore the utmost case is necessary to avoid altering the size distribution when sampling.

References:

East European Method-Book ISSS. Komm. I, VEB—Deutscher Landwirtschaftsverlag, Berlin (1968).

West European methods for Soil Structure Determination. Faculty of Agricultural Sciences, State University Ghent, Belgium (1967).

4.4 Water content

4.4.1. Introduction
The water content is a measure of the amount of water per unit soil mass or soil volume. The moisture content can be expressed:-

on dry mass "w"

$$w = \frac{m_1}{m_2} \qquad (M\,M^{-1})$$

where: m_1 = mass of water
m_2 = mass of dry soil (after drying at 105°C)

on volume basis "θ"

$$\theta = \frac{V_1}{V_b} \qquad (L^3\,L^{-3})$$

where: V_1 = volume of water in sample
V_b = bulk volume of soil sample

The moisture content is also frequently expressed as:
—a percentage of the dry mass of soil (w . 100)
—a percentage of the bulk volume of soil (θ . 100)

The following reference methods of measuring the water content of a soil will be described:
1. Gravimetric method
2. Neutron-scattering method
3. Gamma-ray attenuation method

4.4.2. Gravimetric method
Is the simplest and most widely used method. It is frequently used for the calibration of other indirect methods as it remains the only direct way of measuring soil water.

Principle:
Disturbed or undisturbed wet samples are weighed, dried to constant weight in an oven at 105°C and reweighed. From the different weight measurements the water content on dry mass basis can be calculated.
    Alternatively by multiplying by the bulk density the results may be expressed on a volume basis.

Equipment:
Sampling material (shovel, spade, auger, soil cores), soil containers with tight-fitting lids, oven desiccator with active desiccant, balance.

Procedure:
Place the sample of soil in weighing metal cans with tight-fitting lids. Weigh the samples immediately or store them in such a way that evaporation is prevented. Place the sample in a drying oven (105°C) with the lid off and dry it to constant weight. Remove the sample from the oven, replacing the cover, and place it in a desiccator containing active desiccant until cool. Weigh it again and also determine the tare weight of the sample container. Compute the water content.

Calculation:
The water content is calculated as follows:
—on dry mass basis:

$$W = \frac{m_{s+w} - m_s}{m_s} \qquad (kg\,kg^{-1})$$

where: m_{s+w} = mass of wet soil (kg)
m_s = mass of dry soil (kg)

31

—on volume basis:

$$\theta = \frac{m_{s+w} - m_s}{\rho_w \cdot V_b} \qquad (m^3 \, m^{-3})$$

where: ρ_w = density of water (1000 kg m^{-3})
V_b = bulk volume of soil sample (m^3)

Knowing the bulk density of the soil (ρ_b), the water content on volume basis can be calculated as follows:

$$\theta = \frac{\rho_b}{\rho_w} \cdot W$$

Time required:
Apart from sampling and transportation time the water content determination takes a period of at least one day since 24 hr is normally necessary for complete oven drying.

Cost:
Equipment and running costs are low.

Accuracy:
It is a direct and reproducable measurement. The oven drying at 105°C is itself arbitrary. Some clays may still contain appreciable amounts of adsorbed water even at 105°C. On the other hand, some soil organic matter may oxidize and decompose at this temperature so that the weight loss may not be due entirely to the evaporation of water.

Advantages:
The main advantage of the direct gravimetric method is its simplicity. Moreover it is cheap and no particular skills are required at any stage of the process. It is used as a reference method for the calibration of other water content determination methods.

Disadvantages:
The main disadvantage of this direct method stems from the impossibility of repeating measurements at the same point, and the consequent need for a large number of samples to reduce the error caused by the variability of soil. On the other hand repeated sampling destroys the experimental area.

It requires a great deal of physical effort and time to collect and test samples and to calculate the water content.

References:
Gardner, W. H., (1965). Water content. *In:* Methods of Soil Analysis, I: (Ed. C. A. Black): 82—127. Agronomy No. 9, Am. Soc. Agron., Madison, Wisconsin.
Reynolds, S. G., (1970). The gravimetric method of soil moisture determination. I. Journal of Hydrology, 258—273.

4.4.3. Neutron scattering

Principle:
The neutron-scattering method is based on the fact that fast moving neutrons emitted by a radio-isotope neutron source are slowed down or moderated by collisions with the nuclei of the soil and can be counted by a detector.

Because the moderation ability of the soil nuclei is small compared with that of hydrogen and since the amount of hydrogen associated with water in the soil is generally much greater than that associated with clay, organic matter or other soil particles, the flux of slow neutrons is proportional to the amount of water in a bulk volume of soil.

Equipment:

The portable neutron moisture gauge consists of a probe unit, which is stored in a special shield to protect the user against both neutrons and gamma rays, and a scaler.

The probe unit contains a fast neutron source, a slow neutron detector and an electronic device. The detector count rate is indicated on a scaler or ratemeter.

Access tubes, in which the probe can be lowered during the measurements and which are left in position as long as measurements are required.

Auger to insert the access tube vertically into the soil.

Procedure:

Before the probe is lowered into the access tube a measurement must be done with the probe unit in the special protection shield or by using water in a container as reference.

After determining the standard count, readings are taken at successive depth intervals. At the end of a series of readings in the access tube a new standard count is made.

Each reading is divided by the mean standard reading to obtain a count ratio and referred to the calibration curve to obtain, volumetrically, the water content.

Calculation:

The calibration of the neutron moisture gauge can be carried out in a number of ways viz. by:

gravimetric in situ sampling:
> It is accomplished by comparing many field measurements by the neutron gauge with gravimetric analysis of representative samples taken from the field sites.

preparation of synthetic samples (containers):
> A number of soil containers are filled as uniformly as possible with one soil containing various known amounts of water covering the range of water content that will be encountered in the field.

mathematical models:
> This consists essentially of devising or accepting a model of the neutron gauge response which can be used to describe the gauge response over the entire range of water contents, soil minerals and dry densities encountered.

Time required:

The measurements can be carried out in a relatively short time. Little calculation is necessary to obtain the water contents once the calibration is available.

Cost:

The equipment is expensive. Costs are increased due to the need for a well trained operator.

Accuracy:

As it is an indirect method, the accuracy will largely be associated with the calibration curve viz:
  measurement of mass basis water content
  bulk density
and also the counting device.

Advantages:

Virtually the same soil sample is capable of being measured repeatedly for changes in water content (non-destructive method).

The time required for measurement is relatively short.

The results of measurement in terms of values of soil water content are immediately available.

Measurements can be taken with relative ease to great depths.

The relatively large volume monitored more truly represents the field soil than very much smaller samples taken for the gravimetric measurement of soil water content in water balance studies.

Disadvantages:

The neutron moisture meter is unsuitable for the detection of water content profile discontinuities and measurements close to the surface.

The gauge responds to total hydrogen content. In actual situations appreciable amount of hydrogen are present in clay and, particularly, in organic matter.

The calibration has to be obtained for each soil and soil layer.

The equipment is quite expensive and strict observance of safety rules is necessary.

Remarks:

The water content of the surface layer can be measured by specially adapted equipment. A surface neutron scattering unit is available when the probe is placed on the soil surface and the fast neutrons radiated downwards into the soil. The backscatter of slow neutrons is measured by a detector in the probe. It is important that the probe has intimate contact with the soil surface.

To obtain the water content volumetrically the bulk density of the soil in the vicinity of the access tube has to be known. Therefore different probes are now available containing simultaneously a fast neutron source for the water content determination and a gamma source allowing the determination of the bulk density together with their sensitive detector.

References:

Bell, J. P., (1976). Neutron Probe Practice, Report No. 19. Nat. Env. Res. Council Inst. of Hydrology. Crowmarch Gifford, Wallingford, OXON, England.

Gardner, W. H., (1965). Soil Moisture. *In:* "Methods of Soil Analysis", I. (Ed. C. A. Black), 82—127. Agronomy No. 9, Am. Soc. Agron., Madison, Wisconsin.

IAEA, (1970). Neutron Moisture Gauges. Technical Reports Series, No. 112, 1—95.

IAEA, (1976). Trace Manual on Crops and Soils. Technical Reports Series No. 171, 1—277.

Wilson, R. G., (1970). Methods of measuring soil moisture. International field year for the Great Lakes. Technical Manual Series, No. 1.

4.4.4. Gamma-ray attenuation

The gamma-ray attenuation method is used mostly in the laboratory although the method has also been adapted for field use.

A. Laboratory

Principle:

A radioactive source —mostly ^{137}Cs — whose emitted gamma-rays are collimated into a thin beam, is placed over one side of a soil column and a scintillation counter is located directly opposite.

The degree to which the beam of monoenergetic gamma-rays is attenuated in passing through a soil depends upon the overall density of the material through which it passes.

The changes in attenuation will only represent changes in water content if the density of the soil remains constant and only the water content changes.

Equipment:

A gamma-ray attenuation apparatus consists mainly of two spatially separated units:

—a source of gamma-rays, with a lead shield (for exposure protection) of proper size and a collimator

—a gamma sensitive detector, e.g. a scintillation counter with collimator, connected to a scaler or rate meter.

Also required is a soil container, a mechanism for positioning the container in the beam and a mechanical scanning system which allows the frame, on which source and detector are fixed, to lower or raise along the soil container. A preset timer is desirable.

Procedure:

The soil column is fixed in the gamma-attenuation equipment. Then scanned step by step at each measuring position. The values are used for subsequent computations of water content.

Calculation:

If the soil container has been filled with oven dry soil, the water content on volume basis (θ) can, after scanning, be calculated using the following equation:

$$\theta = - \frac{\log (N_w/N_d)}{0.4343 \, \mu_w \, . \, d \, . \, \rho_w} \tag{1}$$

where:
N_d = the radiation flux transmitted through the container and oven dry soil
N_w = the radiation flux transmitted through the container and wet soil
μ_w = the mass attenuation coefficient of water
d = the thickness of soil through which the collimated gamma beam passes
ρ_w = the density of water

If the soil container has been filled with a soil at a uniform water content (e.g. air dry) other than oven dry the soil water content on volume basis can be calculated using the following equation:

$$\theta = \theta_{ad} - \frac{\log (N_w/N_{ad})}{0.4343 \, \mu_w \, . \, d \, . \, \rho_w} \tag{2}$$

where:
θ_{ad} = the initial water content of the soil determined gravimetrically (e.g. air dry)
N_{ad} = the radiation flux transmitted through the container and air dry soil.

To solve equation (1) or (2):
—the mass attenuation coefficient of water can be determined experimentally or a reasonable estimate can be made from tables for particular gamma-ray energies.
—thickness of the soil column has to be measured at each measuring point.

Time required:
The measuring time is very short (1 minute or less) and the water content can easily be calculated.

Cost:
Very expensive. The setting-up of the experiment and the measurements have to be carried out by a well trained operator.

Accuracy:
The accuracy of the method appears to be about 0.5 to 1.0% by volume.

Advantages:
It is an easy method for monitoring transient water content distributions in soil columns.
   Non-destructive, fast and reproducible method.

Disadvantages:
Expensive. Strict attention has to be paid to all safety rules. Can not be used for soils with important changes in bulk densities during wetting or drying processes.

B. Field
The double-probe gamma-ray method has also been adapted to field use. Two access tubes, one for the source and the second for the scintillation counter, are placed in the soil so that they remain equidistant at all depths.
   Source and detector are both held in an access tube by a rod calibrated for depth and connected to a frame so that they are at the same depth.
   For the rest see "Laboratory method".

Remarks:

In principle this technique offers several advantages over the neutron moisture technique in that it allows much better depth resolution in measurement of soil water content profiles.

Discontinuities between profile layers as well as movement of wetting fronts and conditions prevailing near the soil surface can be detected.

However, this field device is somewhat cumbersome for general usage due to problems related to:
—the accurate installation and alignment of the two access tubes which must be strictly parallel
—determination of the soil bulk density as it might vary in depth and time.

The technique can also be used for the in situ determination of the bulk density (see "Measurement of Soil Density in Situ by Gamma-ray Attenuation").

References:

Gardner, W. H., (1965). Water content. *In:* "Methods of Soil Analysis," I. (Ed. C. A. Black), Agronomy No. 9, Am. Soc. Agron., Madison, Wisconsin.

IAEA, (1976). Tracer Manual on Crops and Soils. Technical Reports, Series No. 171.

Reginato, J. R. and C. H. M. Van Bavel, (1964). Soil water measurement with gamma attenuation. Soil Sci. Soc. Am., Vol. 23, No. 6, 721—724.

Ryhiner, A. H. and J. Pan Kow, (1969). Soil moisture measurement by the gamma transmission method. Journal of Hydrology 9, 194—205.

Wilson, R. G., (1970). Methof of measuring soil moisture. International field year for the Great Lakes. Technical Manual Series, No. 1.

4.5 Soil water retention properties

4.5.1 Introduction

As no satisfactory theory exists for the prediction of the matric potential versus soil water content relationship from basic soil properties, the relationship has to be determined experimentally. The relationship extends from 0 m water head or 0 bar (saturated soil) to approximately -10^5 m water head or -10^4 bar (oven dry soil). Formerly a logarithmic scale was often used in which the symbol pF equals the ^{10}log of the negative of the matric potential when the latter is expressed per unit weight of water and having as dimension cm.

In general the relationship is established by equilibrating a saturated soil sample at a succession of known matric potentials and at each time determining the amount of water that is retained. Plotting the equilibrium water content against the corresponding matric potential gives what is known as the soil water characteristic curve.

It is imperative to use undisturbed soil samples for the high matric potential range (0 to -1 bar), since the structure is of influence on the water retaining properties. For the lower matric potential ($<$ -1 bar or $< -10^3$hPa or < -10 m water head or $>$pF 3 disturbed samples (<2 mm) may generally be used.

If the soil air is kept at atmospheric pressure and the pressure difference across the porous medium is controlled either by a hanging water column or by vacuum the minimum value is limited to -1 bar (Method 1, 2 and 3).

Matric potentials considerably lower than -1 bar can be obtained by increasing the pressure of the air (Method 2).

The soil water characteristic curve, or a more descriptive name for it, the soil water retentivity curve, is used as follows:
—to determine an index of available water in the soil and to classify soil accordingly (e.g. for irrigation purposes).
—to determine the drainable pore space for drainage design.
—to check changes in the structure of a soil (e.g. caused by tillage, addition of organic matter or synthetic soil conditioners).
—to ascertain the relationship between the matric potential and other soil physical properties (e.g. hydraulic conductivity, thermal conductivity).
—to give a general assessment of pore size distribution.

It is recommended that in preparing water characteristic curves, the amount of soil water related to some standard values of the matric potential should always be included viz. at:

pF	Matric potential		
	bar	m(head)	hPa
1	−0.01	− 0.1	− 10
1.78	−0.06	− 0.6	− 60
2	−0.1	− 1.0	− 100
2.5	−1/3	− 3.4	− 333
3	−1	− 10.2	− 1000
4.2	−15	−153	-15.10^3

Descriptive information about the soil pore system—equivalent pore size distribution—can be obtained using the law of capillary rise:

$$h = \frac{2\gamma\cos . \theta}{\rho gr}$$

where: h = height of water rise in a capillary tube (m)
r = radius of the capillary tube (m)
γ = surface tension of the liquid (kg s^{-2})
ρ = density of the liquid (kg m^{-3})
g = acceleration due to gravity (m s^{-2})
θ = contact angle between water and the material composing the capillary and assumed to be zero.

The volume of water extracted from an initially saturated soil sample by a given negative or positive pressure h (dimension length) is equal to the volume of pores having an effective radius greater than the radius r corresponding to the selected value of h in the capillary equation.

The soil pores which differ in size and shape as a result of textural and structural arrangement and estimated from the soil characteristic curve are not only difficult to be classified but the results can be erroneous e.g. in clayey soils due to shrinkage.

If an adequate fitting of the measurement points (h — θ relationship) is possible the pore size distribution can be obtained by mathematical integration.

However storage, availability and transport of soil solution and soil air largely depend on how the total pore space is partitioned and therefore are sometimes classified in:
—macropores (or drainage pores): they affect aeration and drainage.
—micropores(or water holding pores): the remaining pores in which the soil solution is retained or moves very slowly and can be taken up by the plant roots.

As the boundary limits for macro- and micropores could be different from case to case no limits are proposed. However if a distinction is made between macro- and micropores the limits should be stated.

4.5.2 Tension method

Principle:
Undisturbed soil samples are saturated with water and successively brought into equilibrium with the applied negative pressure to a water saturated porous medium on which the samples are placed in close contact. At equilibrium the water content is measured by weighing and related to the applied negative pressure or matric potential.

Equipment:
Sintered glass Büchner funnels connected to an overflow system by means of flexible tubing. Sampling cylinders, oven and balance. The plate in the funnel should be of sufficiently fine porosity to preclude air-entry over the range of negative pressure of concern.

Procedure:
Place undisturbed soil sample on the saturated sintered glass and ensure good contact between the sample and the porous medium. The soil is then fully wetted (saturated) from underneath by raising the point of outflow of the flexible water hanging column. After saturation, cover the Büchner funnel to prevent evaporation and lower the point of outflow until the desired matric potential is obtained.

At each equilibrium, when water flow through the outflow tube of the hanging column ceases, determine the water content gravimetrically.

Calculation:
Subtract the mass of oven-dry (105°C) soil from the mass of wet soil thus giving the equilibrium value "w" of water content per unit mass of oven dry soil at each matric potential level.

$$w = \frac{m_{s+w} - m_s}{m_s} \qquad (kg\ kg^{-1})$$

where: m_{s+w} = mass of wet soil (kg)
m_s = mass of oven dry soil (kg)

The water content can be converted to volume basis (θ), knowing the bulk density of the soil sample, using the following relation:

$$\theta = \frac{w\rho_b}{\rho_w} \qquad (m^3\ m^{-3})$$

where: ρ_b = bulk density of the soil (kg m^{-3})
ρ_w = density of water (1000 kg m^{-3})

Time required:
Depending on the composition and size of the sample and the applied negative pressure. In general it requires from 2 to 14 days to reach equilibrium.

Cost:
Material costs are low. Running of samples in laboratory is not expensive. Highly qualified personnel are not needed.

Accuracy:
Good since the matric potential and water content are directly and accurately determined. The accuracy of the water content by volume depends on the accuracy of the bulk density value.

Advantages:
Technically simple, inexpensive and well suited for routine analysis.

Disadvantages:
Only one sample can be treated per Büchner funnel. It is a desorption curve and the range is limited between saturation and a negative pressure related to the air-entry value of the sintered glass (approximately—0.5 bar or — 500 cm H$_2$O column).

Remarks:
Instead of a sintered glass plate sand can be used as a porous medium. The laboratory set-up consists of a sandbox with a drainage system, connected with an adjustable levelling bottle to realize the

desired negative pressure. A great number of samples can be treated at the same time. The range is limited between saturation and a negative pressure of —0.1 bar.

If the porous medium consists of a mixture of sand and kaolin clay the range can be extended up to —0.5 bar.

References:

Stackman, W. P., G. V. Valk, and G. G. Van der Harst, (1969). Determination of soil moisture retention curves I. Sandbox apparatus. Range pF 0 to pF 2.7. ICW, Wageningen, The Netherlands.

Vomocil, J. A., (1965). Porosity. *In:* "Methods of Soil Analysis", I. (Ed. C. A. Black). Agronomy No. 9, Am. Soc. Agron., Madison, Wisconsin.

West-European Methods for Soil Structure Determination, (1967). Determination of the moisture retention curve up to pF 2.7 with the sandbox method. V, 53, Ghent, Belgium.

4.5.3 Gas pressure method

Principle:
Soil samples on a porous medium (membrane) are saturated with water and successively brought into equilibrium with an applied positive pressure. At equilibrium the water content is determined and related to the applied positive pressure or matric potential.

Equipment:
The pressure apparatus, consists of a pressure chamber containing a porous ceramic membrane or a cellulose acetate membrane, a water delivery tube, an air pressure connection, a regulated source of compressed air. Balance and oven.

Procedure:
The measurement procedure begins with saturating the porous membrane, then the soil samples are placed on it, saturated with water and allowed to stand. Connect the membrane to the outflow system. Close the lid of the extractor. Compressed air or nitrogen is introduced into the chamber at the required pressure, causing water to flow from the soil through the porous membrane until equilibrium is reached.

Once equilibrium is reached—when readings on the outflow burette indicate that liquid water outflow has ceased—release the pressure in the chamber, open the chamber and transfer the samples in metal cans for gravimetric water content determination.

Calculation:
See method 1.

Time required:
The time necessary for establishing equilibrium is dependent upon the sample size, soil type and applied pressure (2—14 days).

Cost:
The price is reasonable taking into account the number of samples which can be treated and the low costs for personnel.

Accuracy:
Good.

Advantages:
Can be used over a wide range of matric potential (0 bar —20 bar), using either undisturbed or sieved soil samples.

Disadvantages:
Due to entrapped air the actual matric potential can be less than that assumed to be applied. Even small air leaks can result in a continuous loss of vapour.

Remarks:
The limit of applied gas pressure is determined by the design of the chamber (i.e. its safe working pressure) and by the maximal air-pressure difference the saturated porous membrane can bear without allowing air to bubble through its pores. Ceramic plates generally do not hold pressures greater than 20 bar, but cellulose acetate membranes can be used with pressures exceeding 100 bar.

If the pressure apparatus is constructed for only one sample, then the air pressure can be increased stepwise. At each step the extracted water is collected and measured. After equilibrium has been reached at the final pressure, the sample is removed from the pressure cell and its water content determined gravimetrically. The water content at various pressure levels can then be calculated by taking into account the amount of outflow at each step.

References:
Chahal, R. S. and R. N. Yong, (1965). Validity of the soil water characteristics determined with the pressurized apparatus. Soil Sci. 99: 98—103.
Hillel, D., (1980). Fundamentals of Soil Physics. Academic Press, 1—413.
Richards, L. A., (1949). Methods of measuring soil moisture tension. Soil Sci. 68: 95—112.
Richards, L. A., (1965). Physical condition of water in soil. *In:* Methods of Soil Analysis I. (Ed. C. A. Black) 128—152. Agronomy No. 9, Am. Soc. Agron., Madison, Wisconsin.
Stackman, W. P., C. V. Valk, and G. G. Van der Harst, (1969). Determination of soil moisture retention curves II. Pressure membrane apparatus. pF 2.7 to 4.2 ICW, Wageningen.
West-European Methods for Soil Structure Determination, (1967). Determination of the moisture characteristic in the range pF 3—4.3 by means of the pressure apparatus according to Richards. V, 62, Ghent, Belgium.
West-European Methods for Soil Structure Determination, (1967). Determination of the moisture characteristic up to pF 3 by means of pressure plate apparatus according to Richards. V, 66, Ghent, Belgium.
Wilson, R. C., (1970). Methods of measuring soil moisture. International field year for the Great Lakes. Technical Manual Series No. 1, 1—19.

4.5.4 Vapour pressure method

Principle:
It is based on the relation between the soil water potential—being the sum of the solute or osmotic potential (h_o) and the matric potential (h_m)—and the water vapour pressure (relative humidity) of the surrounding atmosphere. The sample under investigation is allowed to reach moisture equilibrium with an atmosphere of a known humidity, transfer of water taking place in the vapour phase. The water content of the sample at equilibrium is determined by weighing.

For non-saline soils the vapour pressure method can be used to establish the relation between the water content and the low matric potential (-10 bar to -10^3 bar).

Equipment:
Humidity chamber (desiccator), water bath. Saturated salt solutions. Balance. Oven. Waterjet pump or vacuum pump.

Procedure:
A (wet) sample is put in a separate wire basket and suspended inside a desiccator containing a saturated salt solution. The desiccator is put in a water bath (20°C) and evacuated with a waterjet pump or vacuum pump.

After three days the basket with soil is removed, weighed quickly and replaced inside the desiccator. The weighings are repeated at 24 hour intervals until a constant weight is obtained.

The water content at equilibrium is determined gravimetrically and converted to volume basis using the bulk density.

Calculation:

The value of the soil water potential for which the water content at equilibrium has been determined can be calculated using the following equation:

$$h_m + h_o = \frac{RT}{Mg}\, \ln P/Po \tag{1}$$

where: h_m = matric potential (m)
h_o = osmotic potential (m)
R = universal gas constant (8.3 J mol^{-1} K^{-1})
T = absolute temperature (K)
M = molecular weight of water (0.018 kg mol^{-1})
g = acceleration due to gravity (9.81 m s^{-2})
P = actual vapour pressure of soil air (any unit)
Po = vapour pressure of saturated air (any unit at same temperature)

P/Po is the relative vapour pressure or (P/Po). 100 being the relative humidity of the atmosphere expressed as percentage.

For non-saline soil the osmotic potential ho may be neglected and the matric potential hm can be calculated from equation (1).

The equation (1) can be transformed in such a way that the following relation can be obtained:

$$pF = 6.502 + \log (2 - \log R.H.)$$

where: R.H. = relative humidity.

Time required:
It requires a long time before equilibrium is reached (5—15 days).

Cost:
Equipment and running costs are cheap.

Accuracy:
Good since the matric potential can be calculated accurately.

Advantages:
Direct measurement and cheap.

Disadvantages:
Accurate temperature control is needed, since humidity conditions are dependent on temperature. Long time is required before data are obtained. Condensation of water vapour on the soil sample could occur due to temperature fluctuations. Weighing should be carried out as fast as possible to avoid possible water loss by evaporation.

References:
Stackman, W. P., (1968). Bepaling van vochtspanning en vochtgehalte van gronden door middel van dampspanningsevenwichten. Meded. III. ICW, Wageningen, The Netherlands.
Stackman, W. P., (1974). Measuring soil moisture. *In:* Drainage principles and Applications. III. Surveys and Investigations. International Institute for Land Reclamation and Improvement, Wageningen, The Netherlands.
West-European Methods for Soil Structure Determination, (1977). Determination of soil moisture percentages in the pF range 4.2 to 6.0 according to the vapour pressure method with controlled relative humidity, V, 81, Gent, Belgium.

4.5.5 In situ determination

Principle:
The relation between the matric potential and the water content is obtained in situ using tensiometers and the neutron scattering method or gravimetric method respectively. The soil water characteristic curve is obtained for the range 0 to —0.8 bar.

Equipment:
Tensiometers and neutron moisture gauge or auger and soil containers with tight fitting lids.

Procedure:
Tensiometers are installed in the field at different depths to estimate the matric potential. Simultaneously the matric potential from tensiometer readings and the water content (see methods for soil water content determination) using the neutron moisture gauge (non-destructive method) or gravimetrically (destructive method) are estimated.

Calculation:
The matric potentials are obtained from tensiometer readings while the water contents depend on the method used (see methods for soil water content determination).

Time required:
After installation of the equipment wait until the tensiometers are in equilibrium with the surrounding soil water. It takes a long time (several weeks) before a soil water characteristic curve in the field is obtained.

Cost:
Variable, depending on the method used for water content determination.

Accuracy:
The matric potential can be determined accurately. The accuracy of the water content will depend on the method used (see methods for water content determination).

Advantages:
The measurements are done in situ and consequently are more representative than laboratory measurements. The soil water characteristic curve is obtained from the same soil volume.

Disadvantages:
The obtained relationship (h — θ) will be different if it is obtained during a wetting or drying cycle due to hysteresis effect.

Due to possible entrapped air, the water contents obtained from direct field measurements will be mostly lower than those from laboratory determination.

Reference:
Vachaud, G., C. Dancette, S. Soko, and J. L. Thony, (1978). Méthodes de caractérisation hydrodynamique in situ d'un sol non saturé. Application à deux types de sol du Sénégal en vue de la détermination des termes de bilan hydrique. Ann. Agron., 29 (1): 1—36.

CHAPTER 5

Water and air flow parameters

5.1 Introduction
In the previous chapter, some methods to be used for relatively static characteristics of the structural state were discussed. In this chapter attention will be paid to some methods that can be used to characterize the flow of water and air in soil. Direct measurement of flow rates is rather difficult from an experimental point of view. Besides, a very wide range of values can occur in any particular soil material because flow rates as such are a function of the pressure or concentration gradients in the system. It is therefore advisable to define flow rates as a function of defined pressure or concentration gradients. This aspect will be further discussed when dealing with the various methods. It should be pointed out that flow theory is thoroughly discussed in many recent soil physical textbooks.

Flow of water
Saturated soil
Permeability of saturated soil is defined in terms of the hydraulic conductivity (K_{sat}) which follows from the law of Darcy:

$$K_{sat} = \frac{q}{\Delta H}$$

In which $K_{sat}(ms^{-1})$ is found as a ratio between a flux q (ms^{-1}) and a gradient of the hydraulic head ΔH (mm^{-1}). Saturation is defined as a state in which the soil-water has zero or positive pressure. Some entrapped air may occur. At saturation, all pores are therefore not necessarily filled with water.

Reference methods for measurement of K_{sat} can be classified in four major types, which use:

Undisturbed samples
Sampling cylinders
Gypsum-encased large columns

Auger holes
Auger holes
Piezometers
Inverse auger-hole method

Specific techniques
Air entry permeameter
Double ring infiltration
Drainflow

Calculations
Based on soil texture and moisture retention data.

Unsaturated soil

Permeability of unsaturated soil is also defined in terms of the hydraulic conductivity (K) which follows from the law of Darcy. Unlike K for saturated soil, which has one characteristic value, K for unsaturated soil (K_{unsat}) is different for different moisture contents (and corresponding pressure heads). K_{unsat} is therefore not represented by one value but by an infinite number of values which are usually shown by curves relating K_{unsat} to either the moisture content (θ) or to the negative pressure head (h). K_{unsat} is defined as a ratio between a flux q (ms^{-1}) and a gradient of the hydraulic head ΔH (mm^{-1}), at defined θ or h-values.

Next to hydraulic conductivity, the diffusity (D) is also used to describe flow in unsaturated soil:

$$D = -K \cdot \frac{dh}{d\theta}$$

where $dh/d\theta$ = shape of the moisture rentention curve. Many methods determine D ($m^2 s^{-1}$), which has to be transformed into K values.

A special category of flow is the so-called: "bypass-flow" which describes downward flow of free water along macropores through an unsaturated soil matrix. This type of flow occurs frequently under field conditions, but is not covered in classical soil physics theory. A method to measure bypass flow is included in this chapter, even though it does not represent unsaturated flow in the usual sense.

Reference methods for K_{unsat} and D can be classified in five major types, which use:

Steady flow systems
Infiltration methods
Two-plate method

Unsteady flow systems
One-step method
Instantaneous-profile method

Specific techniques
Sorptivity measurement
Hot-air method

Bypass-flow
Column-outflow method

Calculations
Based on moisture retention data
Calculations based on soil morphology are discussed in Chapter 7.

Flow of soil air

Transport of air in soil in the gaseous phase takes place by diffusion and by mass flow. Further, dissolved molecules of gas are transported by the same mechanisms in soil water. Characteristic constants to describe these processes can be the coefficient of diffusion of a gas in the soil, the Darcy-coefficient of air permeability by mass flow and finally ODR, the oxygen diffusion rate in the bulk soil. Soil air studies also can include observation of the in situ actually found concentrations of different gases. The content of for instance molecular oxygen will be the result of conditions for transport, i.e. soil structure, as well as the action of biological sinks for consumption of oxygen which is highly temperature dependent. A complete description of the physical conditions for transport of air in soil therefore should take into account all of the mentioned parameters. Even though not all of these are determined by the structure of the soil matrix alone, it can be argued, that a survey of soil structure methodology should include methods for determination of them all. Nevertheless, in this paper only methods for transport in the gaseous phase by diffusion and mass flow will be considered.

44

Diffusion

Conditions for diffusion of gases in soil are expressed by the ratio D_{soil}/D_{air}, relating the apparent coefficient of diffusion in soil to the diffusion coefficient of the same gas in air. Diffusion coefficient, D, is the Fick-constant relating diffusion rate and concentration gradient. Four methods for determination of the characteristic constant, D_{soil}/D_{air}, are described in this chapter. All of the methods are laboratory procedures applied to samples of soil. Often the samples constitute "undisturbed" soil taken out in a metal cylinder in the field.

The methods are classified into four groups according to different boundary conditions for the diffusive process. These differences are reflected in measurement techniques, giving practical advantages and disadvantages. The figure beneath summarizes the measurement principles in the methods.

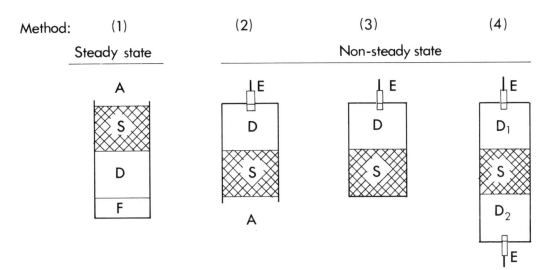

Principles in methods for gas diffusivity determination.
A = atmosphere, S = sample, D = diffusion chamber, E = gas sensor, F = volatilizing liquid.

Permeability

Gas transport by mass flow due to gradients in the overall pressure is simular to the process of water flow in a saturated soil. The permeability, K, is determined from the law of Darcy relating flow rate and pressure gradient.

Two methods are described as reference methods. Both of these can be used for laboratory as well as for field studies.

5.2 Hydraulic conductivity in saturated soil

5.2.1 Sampling cylinders

Principle:

An undisturbed soil sample is collected in a rigid metal or plastic cylinder. The sample is slowly saturated and a water flux is induced through the sample, usually by shallow, constant ponding on top and by allowing free outflow at the bottom. The measured flux is used to calculate K_{sat} by the law of Darcy.

Equipment needed:

Rigid metal or plastic cylinders often with an internal volume of 100 cm^3 or sometimes 300 cm^3; sampling equipment such as spades and devices to push cylinders into soil; laboratory set-up for measurements.

Procedure:
Two possibilities exist: (1) Shallow, constant ponding and monitoring of outflow, and (2) Measurement of the rate of decline of the ponded water at a certain height in a collar on top of the sample, after the addition of water is stopped (Klute, 1965).

Calculations:
For the first procedure K is calculated as follows:

$$K = \frac{Q}{A} \cdot \frac{L}{\Delta H} \qquad (1)$$

For the second procedure:

$$K = \frac{A^1 L}{A(t_2 - t_1)} \cdot \ln \Delta H_1 / \Delta H_2 \qquad (2)$$

where:
Q	=	volume of water per unit time ($m^3 \ s^{-1}$)
A	=	surface area of soil sample (m^2)
A^1	=	surface area of collar on top of sample (m^2) which is smaller than A, to allow for more accurate measurements
L	=	length of soil sample
ΔH	=	gradient of the hydraulic head (=h + z)m (h = pressure head) (also expressed for times t_1 and t_2 for procedure 2)
z	=	height above the bottom of the soil (m)

Time required:
Sampling, and measurement take little time. Batch treatment is possible. Saturation of samples may take several days when they are initially dry, but sampling in clayey soils should occur only when the soil is wet or moist. Then, saturation takes little time.

Cost:
Limited. Cylinders and the laboratory set-up are simple and cheap.

Accuracy:
Good, assuming that undisturbed samples that do not swell upon wetting are obtained. Measurement under controlled laboratory conditions in terms of temperature is favourable, but transportation may induce disturbance and dessication of samples.

Advantages:
Simple method which allows many multiple measurements in a short time. By changing sampling direction, both vertical and horizontal samples may be obtained.

Disadvantages:
Disturbance of the soil may occur when sampling. Sample sizes may be too small to be representative of the soil structure to be characterized. Sampling in stony soils is impossible. Sampling should be confined to soil above the water-table, unless the water-table is lowered before sampling.

Remarks:
Use of small, standard cylinder sizes of 100 cm³ should be avoided as they do not produce representative samples for many soils, particularly those containing silt and clay. Sample size should be a function of the size of the structural elements ("peds") and of the pattern of occurrence of large pores. Maximum cylinder sizes are approximately 6 litres. Samples should be taken in soils that are very moist or wet at the time of sampling. The salt concentration and the temperature of the percolating water should correspond with natural soil conditions.

Reference:
Klute, A. (1965). Laboratory measurement of hydraulic conductivity of saturated soil. *In:* Methods of soil analysis 1. Ed. C. A. Black, ASA monograph 9.

5.2.2 Large encased columns

Principle:
An undisturbed cylindrical column with a diameter and height of 30 cm or more is carefully carved out in situ in very moist or wet soil. An infiltrometer is placed on top. The column is encased in gypsum around the column. The column of soil is saturated and the steady flux of water is monitored during constant, shallow ponding, induced by a mariotte device, and free outflow at the bottom. This flux is equal to K_{sat} because the gradient of the hydraulic head $H = 1$ mm^{-1} (Bouma, 1977).

Equipment needed:
Cylinder infiltrometer with mariotte device; field equipment such as spades and knifes to carve out the column; gypsum or other materials that allow formation of a rigid, well-fitting encasement.

Procedure:
Steady inflow into the column can be measured with a mariotte device, or steady outflow can be collected.

Calculations:
K_{sat} is equal to the measured flux.

Time required:
At least three columns can be carved out per day in surface soil. Sampling in subsoil takes more time. Measurements take little time because fluxes become steady very rapidly in initially very moist or wet soil. All measurements take place in the field; most of the time required is for excavation and sample preparation.

Cost:
Limited. Cylinder infiltrometers are the most expensive part of the equipment.

Accuracy:
Very good, considering presence of a large volume of undisturbed soil which is not transported. The measured flux is equal to K_{sat}, which is therefore obtained in a very accurate manner.

Advantages:
Attractive method to obtain large, undisturbed sample. Method can be used in stony soil.

Disadvantages:
Carving of a column of soil in very loose structures is difficult. Can only be used in soil above the water-table unless the water-table is lowered before sampling. Only K_{sat}-vertical can be determined (see remarks).

Remarks:
Columns should be carved out in soils that are very moist or wet at the time of sampling. The salt concentration and the temperature of the percolating water should correspond with natural soil conditions.
 Two variants of the method have been developed:
(1) The cube method, which not only measures $K_{sat\text{-vert.}}$ but also $K_{sat\text{-hor.}}$. This is accomplished by turning a gypsum covered cube of soil and by making measurements twice. For further information the reader is referred to Bouma and Dekker (1981).
(2) The drain-cube method, which measures K_{sat} in soil directly above and below tile drains. This is done by carving out a cube of soil around the tile drain and by cutting the tile. For further details the reader is referred to Bouma et. al (1981).

Columns and cubes can be strengthened by adding a thin layer of quick-setting cement before application of the gypsum.

References:

Bouma, J. 1977: Soil survey and the study of water in unsaturated soil. Soil Survey paper 13. Soil Survey Institute, Wageningen, 107 p.

Bouma, J. and L. W. Dekker (1981): A method for measuring the vertical and horizontal K_{sat} of clay soils with macropores. Soil Sci. Soc. Amer. Journ. 3: 66.

Bouma, J., J. H. van Hoorn and G. H. Stoffelsen. 1981. The hydraulic conductivity of soil adjacent to tile drains in some heavy clay soils in the Netherlands. J. of Hydrology 50: 371-381.

5.2.3 Auger holes

Principle:
An auger hole is made to a depth well below the water-table. The hole is pumped out and the velocity with which the water moves back into the hole is measured. This velocity is related to K_{sat} with a calculation procedure. A variant of the method removes water to a constant level and maintains it there. The necessary pumping rate is monitored and is used to calculate K_{sat}.

Equipment needed:
Auger, stopwatch, measuring equipment for water levels, pump.

Procedure:
Two procedures can be followed: (1) The rate at which the water-level moves back to its original level before pumping and (2) The pumping rate, which is necessary to maintain the water level at a specified level which is lower than the level in surrounding soil, is measured.

Calculations:
For the first procedure, K_{sat} is calculated as follows (Van Beers, 1978):

$$K_{sat} = C \cdot \frac{\Delta Y}{\Delta t} \tag{1}$$

where: Y = distance between the water-table level and the level in the augerhole (m),
t = time after emptying of the augerhole
C = geometry factor (from tables).

For the second procedure (Zanger, 1953):

$$K_{sat} = \frac{Q}{C.r} \cdot \frac{G}{(G^2 - y^2)} \tag{2}$$

where: Q = flow rate ($m^3\ s^{-1}$)
r = radius augerhole (m)
G = depth augerhole below watertable (m)
y = height water table in the augerhole (m)
C = geometry factor, as a function of y, r and the distance to an impermeable layer, if present.

Time required:
Boring of hole and pumping takes little time. Consistency between replicate measurements can usually be obtained after several hours in slowly permeable soils. A shorter period is usually needed in more permeable soils. Calculations take very little time due to the use of nomograms.

Cost:
Very limited. However, use of electric pumps is expensive and complicated in the field. Procedure 1 can be executed with simple manual pumping. This is more difficult with procedure 2.

Accuracy:
Good in reasonably homogeneous sandy soils, without layering. In layered clayey soils conditions are quite different from those that are assumed in the models that are used to calculate the geometry factor C. Accuracy may be low under those conditions.

Advantages:
Simple field methods which allow many measurements in a relatively short period of time, if applied in sandy, homogeneous soils.

Disadvantages:
Methods can only be used in soil below the water-table. Augerholes in clayey soils will often have puddled walls which do not allow flow of water into the augerholes. Thus, unrepresentative (low) K_{sat} values are obtained. The method is not suitable for unstable soil structures in which bore-holes will collapse. The method yields a K_{sat} that is an undefined mixture of vertical and horizontal flow, although the horizontal component dominates.

Remarks:
Augerholes can be made in a dry period of the year to be used later when the water-table has higher levels. Thus, puddling of sidewalls is avoided. Puddling is difficult to remove by mechanical means once it has occurred.

References:
Beers, W. J. F. van (1958): The auger-hole method. Int. Institute for land Reclamation and Improvement. Bull. No. 1. Wageningen.
Black, C. A. (Ed.) (1965): Methods of Soil Analysis. Agronomy No. 9, American Society of Agronomy Monographs. Madison, Wisconsin.
Zanger, C. N. (1953): Theory and problems of water percolation. U.S. Bur. of Reclamation Eng. Monogr. 8.

5.2.4 Piezometers

Principle:
A closed, open-ended pipe is installed in the soil to well below the water-table and a cylindrical hole is bored beyond the end of the pipe by inserting a small auger into the pipe. Water is pumped out and the velocity with which the water moves back into the hole is measured. This velocity is related to K_{sat} with a calculation procedure. The four-holes method also uses piezometers and can be described as follows:
Two auger holes are made to a depth well below the water-table. Water is pumped from one hole into the other until a stationary condition obtains. At least two piezometers are used to measure the head-gradients between the two holes. The pumping rate is related to K_{sat} with a calculation procedure.

Equipment needed:
Piezometers, small augers, pump, stopwatch, measuring equipment for water levels.

Procedure:
The rise of the water-level in a piezometer is measured after removal of water that was present at hydraulic equilibrium. The piezometers in the four-hole method indicate steady pressures that correspond with the steady flow regime between the two auger-holes.

Calculations:

K_{sat} is calculated as follows (Luthin and Kirkham, 1949):

$$K_{sat} = \frac{\pi r^2}{C(t_2 - t_1)} \ln \frac{Y_1}{Y_2} \qquad (1)$$

where:
- r = radius of piezometer (m)
- t_1, t_2 = time (s)
- Y_1, Y_2 = distance between water-level in piezometer and groundwater level at times t_1 and t_2
- C = geometry factor (m)

The four-holes method uses the following calculation (Snell and Van Schilfgaarde, 1964):

$$K_{sat} = 0.221 \frac{C.Q}{G.Y} \text{ for } d = \frac{1}{3}D \qquad (2)$$

where:
- Q = pumping speed (m³ s⁻¹)
- G = distance between water-table and lower boundary augerholes (m)
- D = distance between the two boreholes (m)
- d = distance between the two piezometers which are placed at regular distances between the two boreholes (m)
- C = geometry factor, which depends on G, d, D and the distance to an impermeable layer, if present. Tables of C values are presented by Snell and Van Schilfgaarde, 1964).

Time required:
Precise placement of the piezometers takes more time than boring a hole, but time required is still limited. The four-holes test is more laborious as it requires installation and monitoring of a set of holes and piezometers.

Calculations:
These take little time due to the use of nomograms and standard tables.

Cost:
Limited. However, use of an electric pump is expensive and complicated in the field. The piezometer method can be executed with a simple handpump. The four-holes method requires use of a pump.

Accuracy:
Good in soils that are reasonably homogeneous with well-defined layers. In layered, clayey soils conditions are quite different from those that are assumed in the models that are used to calculate geometry factors C. Accuracy may be low under those conditions.

Advantages:
Relatively simple method that allows many measurements in a short period of time. The piezometer allows characterization of specific layers in the soil because water can enter the piezometer only at the bottom. This makes the measurement more specific than the one with the augerhole-method (5.2.3). The four-holes method characterizes a large volume of soil.

Disadvantages:
Methods can only be used in soil below the water-table. Piezometer holes in clayey soils will often have puddled sidewalls which do not allow flow of water. This is the more serious since the cavity below the piezometer is very small. Thus, unrepresentative (low) K_{sat} values are obtained. The

method cannot be used in unstable soil materials. The measured K_{sat} reflects only properties in the horizontal direction.

Remarks:
Installation of piezometers in a dry period with low watertables may avoid the puddling problem. Installation of the piezometers should occur with care to avoid flow of water along their shafts. This mechanism is a potential source of error.

References:
Luthin, J. N., and Don Kirkham (1949): A piezometer method for measuring permeability of soil in situ below a water table. Soil Sci. 68: 349-358.
Snell, A. W. and J. van Schilfgaarde (1964): Four well method of measuring hydraulic conductivity in saturated soils. Trans. Am. Soc. Agr. Eng. 7: 83-87, 91.

5.2.5 Inverse auger-hole

Principle:
An augerhole is made to the required depth in soil above the water-table. The hole is filled with water and, after a certain period, the infiltration rate is measured. This rate is "translated" into a K_{sat} value.

Equipment needed:
Auger, stopwatch, measuring tape.

Procedure:
Two procedures can be followed: (1) The rate at which the water level in the borehole moves downwards is measured, or (2) The rate at which water has to be added to the borehole to maintain the water at a constant level is measured.

Calculations:
The following equation is used for the second procedure; which assumes hydraulic gradients of 1 m per m.

$$K_{sat} = \frac{1.15r}{t_2 - t_1} \cdot \log \frac{y_1 + \frac{1}{2}r}{y_2 + \frac{1}{2}r}$$

where: r = radius borehole (m)
$y_{1,2}$ = height of water-level in the borehole above its bottom at times t_1 and t_2 (m)
$t_{1,2}$ = time (s)

Time required:
Soaking of soil is necessary when initial conditions are moist or dry. This may take at least a day. The measurement itself, to be executed several consecutive times to establish consistency, takes several hours.

Cost:
Limited. Equipment is simple.

Accuracy:
Good in reasonably homogeneous sandy soils without layering. In cracked clays, the method is unreliable because natural swelling effects cannot be achieved by a relatively short soaking period. The measured K is an undefined mixture of vertical and horizontal components. Accuracy is reduced by the fact that the hydraulic gradient is often not 1 m m^{-1}.

Advantages:
Simple method which allows many replicated measurements.

Disadvantages:
Cannot be used in soils with very high K_{sat} values and in clay soils.

Remarks:
This method is widely used in the USA as the so-called "percolation test" which is used to estimate soil permeability for spetic tank disposal fields.

References:

Boersma, L. (1965): Field measurement of hydraulic conductivity above a water table. *In:* C. A. Black (Ed.): Methods of soils analysis. I. Agronomy 9. Am. Soc. of Agronomy, Madison Wis. USA: 234-252.

Int. Inst. for Land Reclam. and Impr. (ILRI) (1974): Drainage principles and applications. III: 292-294. Wageningen, The Netherlands.

5.2.6 Air entry permeameter

Principle:
Water infiltrates into dry soil from an infiltrometer, using a high head of water. Once the water has infiltrated to a depth of about 10 cm, water application is stopped. Continued infiltration into the soil causes the development of a negative pressure in the (closed) infiltrometer. This pressure is measured with a manometer, and decreases to a critical value when the air-entry pressure of the wetted soil is exceeded and air bubbles through the wetted soil into the infiltrometer. Then, the pressure being registered by the manometer drops to zero. The critical pressure, the depth of wetting and the infiltration rate are used to calculate K_{sat}.

Equipment needed:
A specially built air-entry permeameter must be available (diameter 20 or 30 cm). No further special equipment is needed.

Procedure:
A horizontal plane is prepared in the soil at the required depth on top of which the air-entry-permeameter is placed. The experimental procedure is described above.

Calculation:
K_{sat} is calculated with the following equation:

$$K_{sat} = \frac{Q}{A} \cdot \frac{z}{(Y - \frac{1}{2} h_a) + z} \tag{1}$$

where: Q/A = flux, just before water application is stopped (ms^{-1})
  z = depth of water front (m)
  Y = height of water in standpipe above the soil surface at the moment when water application is stopped (m)

The term h_a is measured as follows:

$$h_a = h_{min} + G + z \tag{2}$$

where: h_a = air entry valve (m)
  G = height manometer above soil surface (m)
  h_{min} = minimum pressure in manometer (m)

Time required:
One measurement can be made in about one hour, including placement of the permeameter. At one location, water can only be applied once.

Cost:
Restricted to the air-permeameter.

Accuracy:
Good in dry soils with well defined air entry values. Poor under other conditions.

Advantages:
Relatively rapid method which uses a large sample.

Disadvantages:
The method can only be used in soil with well defined air-entry values. occurrence of continuous macropores results in instant air breakthrough, making the test useless. Application can only be considered in sandy, nonswelling and non-stony soil. K_{sat} of clayey soils can only be measured when the soil is wet and swollen, not in dry condition.

References:
Bouwer, H. (1966): Rapid field measurement of air entry value and hydraulic conductivity of soil as significant parameters in flow system analysis. Water Res. Res. 2, 4: 729-738.
Topp, G. C. and M. R. Binns (1976): Field measurement of hydraulic conductivity with a modified air-entry permeameter. Can. J. Soil Sci. 56: 139-147.

5.2.7 Double ring infiltration

Principle:
Two concentric cylinders are placed on top of the soil and they are filled with water. The water levels in the two rings are maintained at a constant level, while the downward rate of movement of the water-level in the inner-ring is observed. This rate is representative for K_{sat} when wetting has progressed beyond a certain depth for an extended period of time. Higher initial infiltration rates can be measured as well.

Equipment needed:
An inner- and outer metal cylinder, to be used as infiltrometer. Measuring equipment to observe the downward rate of movement of water in the inner cylinder.

Procedure:
The infiltration rate is observed until constant. A bottle with mariotte device can be used to measure the infiltration rate.

Calculation:
Vertical infiltration in a deep, homogeneous soil is characterized by:

$$K_{sat} = \frac{Q}{A} \cdot \frac{z}{(h - h_f) + z} \tag{1}$$

where: Q/A = flux (ms^{-1}). (Q = flow rate in $m^3 s^{-1}$ and A = surface area infiltrometer in m^2)
h = pressure head at soil surface (m)
h_f = pressure head at the infiltration front (m)
z = depth infiltration front below surface (m)

As the infiltration front moves deeper, (h—hf) will become much smaller than 2. Thus, equation (1) changes to:

$$K = \frac{Q}{A} \qquad (2)$$

Time required:
Varies in different soils, but usually several hours per measurement. Steady infiltration rates are directly equal to K_{sat}. No additional calculations needed.

Cost:
Limited, only infiltrometers and measuring equipment are required.

Accuracy:
High, when fluxes are accurately measured. Inner infiltrometers often have a diameter of 30 cm, yielding relatively large samples.

Advantages:
Simple, direct method. Steady infiltration rates are equal to K_{sat} in homogeneous soils.

Disadvantages:
The method does not work in soils with slowly or very highly permeable horizons that occur relatively close to the soil surface. Then, the pressure head gradient is most likely *not* equal to 1 mm [1] with the result that the infiltration rate is lower than K_{sat}. Also, lateral flow of water may occur away from the outer ring in soils with macropores. This disrupts the required one-dimensional system.

Remarks:
K_{sat} in swelling clayey soils can only be measured in initially wet soil. The method being described here can be used then. However, continuous downward flow in the soil should be assured.

References:
FAO (1979). Soil Survey Investigations for irrigation. Rome.
Winger, R. J. (1960). In place permeability tests and their use in subsurface drainage. Int. Congr. Comm. Irrig. and Drainage. Madrid.

5.2.8 Drainflow

Principle:
Water-table levels and drain discharges are measured during a certain period. K_{sat} is calculated with an equation which expresses relationships between drain discharge, water-table level and soil profile characteristics.

Equipment needed:
Bucket, stopwatch, piezometers, and measuring tape.

Calculation:
The calculation is based on the drainage equation of Hooghoudt, as follows:

$$q = \frac{8 K_o \, d.h}{L^2} + \frac{4 K_b \, h^2}{L^2}$$

where: q   = drain discharge per m^2 of surface area (ms^{-1})
        h   = height of water-level between the drains, measured halfway between
               the drains (m)
        d   = equivalent thickness of the layer through which flow occurs, to be
               determined from tables (m)
        L   = distance between the drains (m)
        K_o = K_{sat} of soil below the drains (ms^{-1})
        K_b = K_{sat} of soil above the drains (ms^{-1})

The equivalent thickness d is smaller than the real thickness of the layer through which flow occurs, as it expresses the effect of the resistance of the soil surrounding the drains. For specific calculation details, the reader is referred to the reference below.

Time required:
The method is laborious, because repeated measurements are needed. Drain discharge depends, of course, on weather conditions.

Cost:
Low, except for time requirements.

Accuracy:
Limited. Flow above the drains is verticle; the direction below the drains is variable. The volume of soil being characterized is very large. The equation being used is valid for "ideal" soil with a uniform profile, constant thickness and permeability and steady flow. This is often not true, and the value of d may therefore vary considerably.

Advantages:
Very large sample consisting of undisturbed soil, except for possible compaction around the drains.

Disadvantages:
Method can only be used when drains discharge water and when the water level in the ditches is below drain level.

Reference:
Institute of land Reclamation and Improvement (ILRI) (1974). Drainage principles and applications Vol. 3: 337-431.

5.3 Hydraulic conductivity in unsaturated soil

5.3.1 Infiltration methods

Principle:
A column or cylinder, filled with undisturbed soil, is subjected to a constant flux of water at its upper surface. After some time a constant moisture content and pressure head is observed below the surface of infiltration. The measured flux is equal to K_{unsat} at the measured water content and pressure head.

Equipment needed:
Column filled with soil; equipment to apply water. Tensiometer and equipment to measure water contents. Required equipment is simple and not expensive.

Procedure:
Three methods are followed to apply water at a constant rate: (1) Water can be sprinkled on the column at a series of rates that are lower than K_{sat}. (2) Water is applied through a porous plate to

which a negative pressure is applied (Youngs, 1963) or (3) water is applied through a crust with a higher resistance which allows a small positive head on top of the crust (the crust method). Crusts are made by mixing quick setting cement and sand in particular mixtures (usually 10%, 20% and 50% cement). Sand and cement are mixed when dry and a paste is formed by adding water. This paste is put on top of the soil and hardens in a few minutes. The infiltration rate through the crust is measured as well as the subcrust pressure head. The latter should preferably be measured at two points to allow an estimate of the verticle hydraulic gradient which is usually 1 m m^{-1} in a deep column of soil. The measurement is repeated with several crusts.

Calculations:
The flux is equal to K$_{unsat}$ at the measured subcrust pressure head when grad H = 1 mm^{-1}. No calculations are needed.

Time required:
Limited. preparation of samples will take some time; but measurements can be made quite rapidly since the method is restricted to pressure heads that are relatively close to saturation, when fluxes are relatively high.

Cost:
Limited.

Accuracy:
Excellent, because flux and associated pressure head and/or moisture content are obtained directly.

Advantage:
Simple, very accurate method which yields K values in a pressure head range for which K values are difficult to obtain with other methods. The method by which the crust is applied in the crust method ensures perfect contact between soil and crust.

Disadvantages:
The range of the method is restricted to pressure heads of approximately 0 to —80 cm, which are associated with fluxes of at least several mm per day. Procedure 1 cannot be used in soil with macropores since bypass flow will occur (see 5.4). In procedure 2 problems may occur with the contact between soil and plate. Procedure 3 (the crust test) is the most attractive.

Remarks:
Earlier versions of the crust method used gypsum-sand crusts. These may dissolve slowly during the measurements. Use of quick setting cement overcomes these problems. The method, as described here, measures inflow into the soil. Outflow from a soil column can also be measured, but then the column should be sufficiently long.

References:
Youngs, E. G., (1963). An infiltration method of measuring the hydraulic conductivity of unsaturated porous materials. Soil Sci. 97: 307-311.
Hillel, D. and W. R. Gardner, (1970). Measurement of unsaturated conductivity and diffusivity by infiltration through an impeding layer. Soil Sci. 109: 149-153.
Bouma, J., C. Belmans, L. W. Dekker and W. J. M. Jeurissen, (1983). Assessing the suitability of soils with macropores for subsurface liquid waste disposal. Journal of Environmental Quality 12 (3): 305-311.
In this paper the latest version of the crust test is described. Earlier references are given in this paper.

5.3.2 **Two-plate method**

Principle:
A soil sample is placed between two porous-plates which are subjected to different (negative) pressure heads. A steady flow is established and K$_{unsat}$ is obtained by also measuring the two pressure heads in soil adjacent to the two plates.

Equipment needed:
Porous plates, tensiometers, flux measuring equipment.

Procedure:
The two plates can be subjected to negative pressure heads (Laliberte and Corey, 1967) or to positive gas pressure (Elrick and Bowman, 1964). A small difference in pressure is induced between the two plates, so as to allow calculation of K at the average pressure head.

Calculations:
K_{unsat} is considered constant for the pressure head range being considered. K_{unsat} for the average pressure head follows from:

$$K_{unsat} = \frac{Q}{A} \cdot \frac{1}{1 - \Delta h / \Delta z}$$

where: Q = volume of water that leaves the sample during time t $(m^3 s^{-1})$
       A = cross sectional surface (m^2)
       h = pressure head (m)
       z = depth below the soil surface in the sample (m)

Time required:
At lower pressure heads, equilibrium steady-state flow systems are only reached after several days, which makes the method very time consuming. At higher pressure heads, measurements take less time, although frequent observations of fluxes and pressure heads are needed.

Cost:
Material costs are low, but labour cost may be high.

Accuracy:
Good, when the gradient of the pressure head is low. However, fluxes can only be measured if the gradient has some value. At certain moisture contents, K may change strongly with small changes in moisture content. Then, the method is rather inaccurate.

Advantages:
Relatively simple method requiring only simple, direct calculations.

Disadvantages:
Labour intensive method. Results are approximate due to the averaging procedure involved. Contact of soil and plates is critical to allow flow, and this may cause problems in swelling or unstable soils because plates should not be pressed too hard onto the soil sample. The closed system does not allow air to escape, nor does it allow air-entry.

Remarks:
A variant of the method has been published by Henseler and Renger (1968) who used a decreasing pressure head difference in a "falling-head" method. Nielsen et al (1960, 1961) used one porous plate at the bottom of the sample to induce a constant flux there, while allowing free evaporation at the top of the sample.

References:

Henseler, K. L. and M. Renger, (1968). Die Bestimmung der Wasserdurchlässigkeit in Wasserungesättigden Boden mit der Doppeltmembran-Druckapparatur. Z. Pflanzenern. Bodenk. 122: 220-228.

Nielsen, D. R., D. Kirkham and E. R. Perrier, (1960). Soil capillary conductivity: Comparison of measured and calculated values. Soil Sci. Soc. Amer. Proc. 24: 157-160.

Nielsen, D. R. and J. W. Biggar, (1961). Measuring capillary conductivity. Soil Sci. 92: 192-193.

Elrick, D. E. and Bowman, D. H., (1964). Improved apparatus for soil moisture flow measurements. Soil Sci. Soc. Amer. Proc. 28: 450-453.

Laliberte, G. E. and A. T. Corey, 1967: Hydraulic properties of disturbed and undisturbed samples. In: Permeability and capillarity of soils. Am. Soc. Test. Mat. STP 417 p. 56-71.

5.3.3 Instantaneous-profile method

Principle:
A soil profile is wetted in situ and covered with a plastic sheet to stop evaporation. Moisture contents and pressure heads are measured at different times during natural drainage of the soil. These data are used to calculate K.

Equipment needed:
Tensiometers, equipment to measure moisture contents (e.g. neutron probe), other miscellaneous equipment.

Procedure:
See Principle.

Calculations:
Fluxes are calculated as follows:

$$q(z, t) = - \int_{o}^{z} \left(\frac{d\theta}{dt}\right)_z \ dz \tag{1}$$

where: $q(z, t)$ = flux at depth z and time t (ms^{-1})
θ = moisture content $(m^3 m^{-3})$
z = depth below surface (m)
t = time (s)

K is calculated as follows:

$$K = \frac{q(z, t)}{1 - \dfrac{dh(z, t)}{dz}} \tag{2}$$

Time required:
The measurement may take several weeks, particularly in clayey soils with low internal drainage rates. Field observations take much time.

Cost:
Relatively high, due to equipment needs and labour intensive procedures.

Accuracy:
Good. Method is directly derived from field drainage experiments and posseses thus a built-in validation.

Advantages:
Method yields representative data for field soils in a pressure head range that is relevant for internal drainage problems.

Disadvantages:
The assumption of one-dimensional flow does not allow for the occurrence of subsurface horizons with different properties. Application is sloping soils is not possible. Method is not suitable to obtain values near saturation because of very rapid drainage. Very time consuming when low fluxes occur.

Remarks:
The method can also be used allowing evaporation at the soil surface. The observed pressure heads allow a determination of the depth of the "zero-flux-plane" from where upward flow occurs towards

the soil surface and downward flow towards the water-table or subsurface horizons. (See Richards et. al., 1956). A similar method for soil samples in the laboratory has been proposed by Wind (1969) and Boels et. al., (1978), see Section 5.3.4.

References:
Hillel, D. I., V. D. Krentos and Y. Stylianon, (1972). Procedure and test of an internal drainage method for measuring soil hydraulic characteristics in situ. Soil Science 114: 395-400. (Provides a specific example).
Richards, L. A., W. R. Gardner and G. Ogata, (1956). Physical processes determining water loss from soil. Soil Science Soc. Amer. Proc. 20: 310-314.

5.3.4 One-step method

Principle:
A soil sample is placed on a porous plate in a pressure chamber or in a Büchner funnel. The sample is in contact through the plate with water of a known pressure. Suddenly the pressure in the chamber is increased to a known value (or the negative pressure below the porous plate is reduced to a known value). Outflow from the sample is measured as a function of time. Once a new hydraulic equilibrium has been reached, the moisture content in the sample is determined. Data are used to calculate the diffusity (D) which can be transformed into K.

Equipment needed:
Pressure chamber or Büchner funnel with associated equipment.

Procedure:
See principle.

Calculations:
The equation used to calculate D is as follows: (Gardner, 1962)

$$D(\theta) = \frac{4L^2}{\pi^2 (\theta_t - \theta_f)} \cdot d\theta_t/dt$$

where: $D(\theta)$ = diffusity (m^2s^{-1})
      θ_t = moisture content at time t (m^3m^{-3})
      θ_f = final moisture content (m^3m^{-3})
      t = time after application of pressure (s)
      L = height soil sample (m)

Time required:
Depending on the pressure range being considered. Equilibrium takes more time to reach at lower pressure heads.

Cost:
Limited; equipment is relatively simple and inexpensive.

Accuracy:
Limited, because of relatively small sample sizes. Outflow volumes can be measured accurately with a graduated burette. Transformation of D into K involves introduction of some error.

Advantages:
Relatively simple method that yields good data for a limited range of pressure heads and moisture contents.

Disadvantages:
Rather time consuming at low fluxes. Unsuitable in swelling soils. Cannot be used to obtain K near saturation, because of rapidly changing outflow rates.

Remarks:
This method is suitable to be applied simultaneously in a series of samples.

References:

Gardner, W. R. (1962). Note on the separation and solution of diffusion type equations. SSSAP 26: 404-405.

Jaynes, D. B. and E. J. Tyler, (1980). Comparison of one-step outflow method to an in situ method for measuring hydraulic conductivity SSSAJ44: 903-907.

Gupta, S. C., D. A. Farrell and W. E. Larson (1974). Determining effective soil water difusivities from one-step outflow experiments. SSSAP 38: 710-716.

Wind, G. P., (1969). Capillary conductivity data estimated by a simple method. Proc. Symp. Water in the Unsaturated Zone, Wageningen, Netherlands.

Boels, D., J. B. H. M. van Gils, G. J. Veerman and K. E. Wit, (1978). Theory and system of automatic determination of soil moisture characteristics and unsaturated hydraulic conductivities. Soil Science 126: 191-199.

5.3.5 Sorptivity measurement

Principle:
Water is applied under pressure to a soil sample through a porous plate. Pressure is applied in such a way that the cumulative adsorption of water is proportionate to $\sqrt{t}$. Sorptivity values obtained are used to calculate the diffusity (D) which can be transformed in K_{unsat}.

Equipment needed:
A special apparatus is needed for applying water to a soil sample with a rate that is proportionate to $\sqrt{t}$. Additional equipment is minimal.

Procedure:
Water is applied with a pump and special gears which allow the cumulative absorption of water (i) to be proportionate to $\sqrt{t}$. Various combinations of pumping velocities, gear size and pump-volume can be used to induce specific Sorptivity (S) values, where:

$$i = S \sqrt{t}$$

The water content and pressure head are measured in soil adjacent to the porous plate, as soon as S is constant. Frequently this requires only a few minutes. Five to ten determinations can be made in each sample.

Calculations:
Diffusity (D) is calculated from the obtained S values:

$$D(\theta_1) = \frac{\bar{\mu} S^2}{4(\theta_1 - \theta_o)} \cdot \left[\frac{\theta_1 - \theta_o}{1 + \gamma \log e} \frac{d}{d\theta_1} \log S^2 - \frac{1 - \gamma}{1 + \gamma} \right]$$

where γ = constant varying between 0.50 and 0.67.

Time required:
Limited. A sample can be used for several determinations and calculations are rapid when a computer program is available.

Cost:
Limited, once the special apparatus has been bought.

Accuracy:
Good, although complex calculations for D are needed, which include the empirical factor γ. This aspect tends to decrease accuracy.

Advantages:
Simple, rapid method that allows multiple measurements in a short time.

Disadvantages:
Complex calculations for D. Additional error introduced when transforming D to K. Applications limited to sandy and silty soils. Swelling soils are more difficult to characterize with this method.

Remarks:
This is a promising well-tested method that deserves to be more widely applied.

Reference:
Dirksen, C. (1979). Flux controlled sorptivity measurements to determine soil hydraulic property functions. Soil Sci. Soc. Amer. J. 43: 827-834.

5.3.6 Hot-air method

Principle:
A moist soil sample is dried at the top for a limited period of time with hot air. The moisture content distribution in the sample after treatment is measured and is used to calculate D.

Equipment needed:
A hot-air gun capable of producing a stream of hot air with a temperature of $\pm 200°C$. A balance for weighing the sample.

Procedure:
Undisturbed soil samples are taken in 10 cm high metal cylinders. After saturation, the samples are desorbed to a pressure head of approximately 30 cm. Hot air is blown onto the samples during a period of approx. 10 minutes. The soil is pushed from the cylinder and is cut in 5 mm thick slices, of which the moisture content is measured. Moisture content is plotted versus depth and this graph is used to calculate D.

Calculations:
Diffusivity is calculated as follows:

$$D(\theta_x) = \frac{1}{2t}\left(\frac{dx}{d\theta}\right)_{\theta_x} \cdot \int_{\theta_x}^{\theta_i} x\, d\theta \qquad (1)$$

where:
D = Diffusivity (m^2s^{-1})
θ_x = moisture content at depth x after t sec
θ_i = initial moisture content
t = duration of evaporation (s)
x = depth (m)

The following boundary conditions apply:
1) the sample should be homogenous
2) the evaporative loss, which is being measured repeatedly during the procedure, should be proportionate to $\sqrt{t}$ and
3) the moisture content in the bottom of the sample should remain constant (semi-infinite sample).

61

Time required:
Little time needed. The measurement takes some 45 minutes all together. Calculations can be made with a computer.

Cost:
Quite limited.

Accuracy:
Reasonable, but there are several sources of error: evaporation of water during sampling; viscosity changes of the water during the measurement and the approximate character of the calculation procedure.

Advantages:
Rapid, simple method which allows multiple measurements in a short time. It provides data for a wide range of moisture conditions.

Remarks:
Some theoretical limitations of the method have to be recognized: Water transport is not isothermal. The method does not work well in sands, probably because of irregular evaporation in a sandy matrix when applying a stream of forced air, but results in clayey soils are good. Moisture retention curve is needed to obtain K$_{unsat}$.

References:

Arya, L. M., D. A. Farrell and G. R. Blake, (1975). A field study of soil water depletion patterns in presence of growing soybean roots. I. Determination of hydraulic properties of the soil. Soil Sci. Soc. Amer. Proc. 39: 424-430.

5.3.7 Calculations based on moisture retention data

Principle:
The method is based on a model-concept in which soil pores are represented as bundles of cylindrical pores. Their (equivalent) size distribution is derived from moisture retention data. When saturated, all pores are filled with water. As the pressure head decreases, increasingly smaller pores are emptied. K decreases accordingly in a pattern that is related to the pore-size distribution.

Equipment needed:
None, if moisture retention data are available. Only computer calculations are needed.

Procedure:
Two procedures are followed. One procedure was initially published by Marshall, Millington and Quirk and was later summarized by Green and Corey (1971). The second procedure was derived by Brooks and Corey (1964).

Calculations:
The first procedure is based on Poisseuille's law and starts with an equation defining the pressure head (h) at which pores with radius r are still filled with water:

$$h = \frac{-2\delta}{\rho.g.r} \tag{1}$$

where: δ = surface tension of water (kg s^{-2})
  ρ = density of water (kg m^{-3})
  g = acceleration of gravity (ms^{-2})
  r = radius cylindrical pore (m)

62

Poisseuille's law relates the flow rate of water to the size of a (cylindrical) pore, as follows:

$$K = \frac{\pi r^2 \rho g}{8\eta} \tag{2}$$

in which: η = viscosity of the water (kg m^{-1} s^{-1}).

Equations (1) and (2) are combined and the moisture-content range of the soil is divided in n-intervals ($i = 1, 2, \ldots n$). The porevolume in an interval i is occupied by cylindrical pores that are still filled with water at the pressure head h_i. h values are read from the moisture retention curve. The following equation results after introduction of a pore-interaction model:

$$K(\theta)_i = K_s/K_{sc} \cdot \frac{30\delta^2}{\rho g \eta} \cdot \frac{f^p}{n^2} \sum_{j=i}^{m} (2j + 1 - 2i)\, h_j^{-2} \tag{3}$$

where $K(\theta)_i = K$ for θ_i ($i = 1$ in saturated soil). $i = m$ = pore-class with lowest water content for which calculations are made. p = constant f = porosity (m^{3}m^{-3}). The term K_s/K_{sc} is the matching factor which defines the ratio of the measured versus the calculated K_{sat}. The second procedure is based on the following equation (4):

$$K(h) = K_s \left(\frac{h_a}{h}\right)^n \tag{4}$$

where h_a = air-entry value. n = measure for pore-size distribution as defined by Brooks and Corey (1964). K_s and h_a must be measured. n is derived from the moisture retention curve. They use the factor S_e which is defined as "effective saturation":

$$S_e = \frac{S - S_r}{1 - S_r}$$

where: S   = ratio of water-filled pores to total porosity and
S_r = residual saturation. They prove that:

$$n = 2 + 3\lambda$$

where: λ = d (log S_e)/d (log h)

Time required:
Quite limited if moisture retention data are available. Measurements of K_{sat} (for matching purposes) and h_a may take considerable time.

Cost:
Very limited if activities can be restricted to calculations.

Accuracy:
Higher for the first procedure than for the second, because of the often poorly defined h_a values. Particular h_a values can only be well determined in well sorted sands. Clayey soils with macropores have undefined h_a values. The first procedure works rather well in sandy soils for which the equivalent pore-size model is suitable. Problems arise in clayey soils as is evident by often very small matching factors, as cited by Green and Corey. The matching factor has no physical significance.

Advantages:
Both procedures are relatively simple and cheap, allowing multiple applications in a short period of time.

Disadvantages:
Both methods have severe theoretical limitations for soils with poorly defined h_a values and with a pore size distribution that cannot be represented by a static capillary-bundle model. Unfortunately, most soils belong to both of these problem categories.

Remarks:
These methods should be applied with utmost care to avoid generation of irrelevant data.

References:
Procedure 1: Green, R. E., and J. C. Corey (1971). Calculation of hydraulic conductivity: A further evaluation of some predictive methods. Soil Sci. Soc. Amer. Proc. 35: 3-8.
Procedure 2: Brooks, R. H. and T. Corey (1964). Hydraulic properties of porous media. Colorado State Univ. Hydr. Papers 3. 27p.

5.4 Bypass flow

5.4.1 Column outflow method

Principle:
A large, undisturbed dry column of a soil with macropores is subjected to sprinkling irrigation with a defined intensity and duration. The mass of the soil is determined. The moment that water first leaves the column is observed and the outflow rate is measured during sprinkling. The column is weighed after termination of the experiment. The increase of the soil mass is attributed to the adsorption of water. The volume of water that left the still unsaturated soil mass is expressed as a percentage of the applied water to obtain a percentage value for bypass flow.

Equipment needed:
Large cylinders of soil. Sewer-pipe segments with a diameter and length of 20 cm have been used successfully. A simple stand to allow collection of outflow from the column. Sprinkling equipment.

Procedure:
See principle.

Calculations:
None. The outflow at a given time is compared with the inflow to calculate the percentage of bypass flow. The determination of the dry mass at the start of the experiment allows an estimate of soil porosity and of the associated moisture content at saturation. According to Darcy flow theory, outflow from the column can only occur when the soil is saturated (or more precise: when pressure head $h = 0$ cm at the bottom of the core and $h = -20$ cm at the top of a 20 cm high core). Outflow during by-pass flow occurs much more rapidly.

Time required:
Little. When commercial sprinkling equipment is used, eight samples can be run in about 30 minutes. The measured outflow data are directly applicable; no additional calculation needed.

Cost:
Very limited.

Accuracy:
Good. the measurement of outflow is accurate. The inflow rate is more difficult to measure when intermittent sprinkling is applied.

64

Advantages:
Simple, cheap and rapid method which allows multiple measurements of an important phenomenon that has so far been ignored.

Remarks:
Free outflow at the bottom of the core is only realistic in soils that have free macropore-drainage to the subsoil. If the macropores, such as cracks, end at a certain depth, they will fill up with water and water will have to slowly infiltrate into the soil matrix. Bypass-flow in a short core, with continuous pores, is then not representative for field conditions. Conditions after filling up of the macropores were described by Bouma and Wösten (1984). Sample size should be such that macropore patterns are well represented. When this is assured, this well tested method produces good results.

References:

Bouma, J., L. W. Dekker and C. J. Muilwijk, (1981). A field method for measuring short-circuiting in clay soils. J. of Hydrol. 52: 347-354.

Bouma, J. and J. H. M. Wösten, (1984). Characterizing ponded infiltration in a dry cracked clay soil. J. of Hydr. 69: 297-304.

Bouma, J. (1985). Using soil morphology to develop measurement methods and simulation techniques for water movement in heavy clay soils. *In:* J. Bouma and P. A. C. Raats (Eds.). Proc. of the ISSS Symposium on water and solute movement in heavy clay soils. ILRI publication 37, Wageningen, Netherlands p. 298-316.

5.5 Air diffusion

5.5.1 Steady state method

Principle:
A steady state of diffusion of gas through a soil sample is established. Diffusion rate is determined as loss of the volatilizing liquid, which is used as diffusing gas. From this value D_{soil} and the ratio D_{soil}/D_{air} can be calculated.

Equipment needed:
Sampling cylinders, size dependent of soil/material to be measured, but equivalent to dimensions of the diffusion apparatus. A reservoir with liquid and a soil sample holder. Balance. Field sampling equipment. (Equipment for potential-controlled drainage of soil samples).

Procedure:
The soil sample is placed in airtight contact with a vessel containing a liquid with high volatility. The loss of gas from the system is followed by weighing through a period of time.

Calculation:
According to Ficks first law, diffusion in one direction for a steady state condition can be described by

$$\frac{dc}{dt} = - D \cdot A \cdot \frac{\Delta c}{l},$$

where:
D = diffusion coefficient
t = time
c = concentration of gas in question
A = cross sectional area of sample
l = sample height

If partial pressures in absolute figures are used instead of concentration gradients the calculation of diffusion coefficient will be

$$C = (\Delta p \ . \ K) \ . \ (D \ . \ \frac{A}{l}) \ <=>$$

$$D = C \ . \ \frac{l}{\Delta p . \ K \ . \ A}$$

where: C = average loss of gas from the system in the period of steady state condition, $g \ s^{-1}$

Δp = vapour pressure of the gas in question at actual temperature

K = a correction factor, including converting the result to standard temperature and pressure (S.T.P.) condition.

K can be calculated as

$$K = \beta_o \ . \ \frac{T}{273} \ . \ \frac{P_o}{P}$$

where: β_o = concentration in $g \ cm^{-3}$ of the gas in a pure state at 273° K and a pressure of 1 mm Hg (β_o = Moleweight/(22414 . 760)),

T = temperature in ^{o}K

P = atmospheric pressure in atm.

P_o = 1 atm

C is calculated from the loss of weight, with correction for water evaporation.
Δp is read from physical tables.
l and A are known constants depending of the apparature used.
Finally D can be related to tabulated values of D_o for diffusion in air to yield the ratio D_{soil}/D_{air}.

Time:
Sampling time-consuming dependent on depth of sampling etc. Measurement takes rather long time ($\sim$ 6 hours), with replicate inspection. Calculations simple.

Costs:
Materials costs are low, running of samples in lab. Not expensive. Highly qualified personnel not needed.

Accuracy:
Poor, ambient temperature and pressure may change during a run. Evaporation of water from sample may change diffusive conditions.

Advantages:
Cheap, simple method.

Disadvantages:
Long tests necessary, implicating changing conditions for the test. Mass flow can occur, caused by different diffusion coefficients for counter diffusing material.

Reference:
Penman, H. L. (1940). Gas and vapour movements in the soil. I. The diffusion of vapours through porous solids. Journal of Agricultural Science, 30, 438-462.

5.5.2 Non-steady state method: open system

Principle:
The sample to be measured is placed in airtight contact with a diffusion chamber, the one surface of the sample facing the free atmosphere. A gradient of concentration of a gass is established by replacing the atmospheric air in the chamber by a pure gas. Diffusion is followed by repeated measurement of gas concentration in the chamber.

Equipment needed:
Sampling cylinders; field sampling equipment. (Equipment for potential-controlled drainage of samples). Sample holder/diffusion chamber. Gas measuring device (different principles in variants of the method). Computer.

Procedure:
The concentration gradient is established by flushing the diffusion chamber with a gas different from the atmosphere. Often pure nitrogen is used, in which case oxygen is the gas detected in the chamber to follow diffusion through the sample.

Calculation:
The non-steady state process of diffusion can be described by the equation

$$\frac{dC}{dt} = \frac{D}{c_a} \cdot \frac{d^2C}{dx^2}$$

where C = concentration
        t = diffusion time
        x = diffusion distance
        c_a = airfilled porosity
        D = diffusion coefficient

Based on an approximate solution of this equation the diffusion coefficient can be calculated according to

$$D = - \frac{c_a}{a_1^2} \cdot K$$

where a_1 = the smallest positive root of $a \, tg(ah_s) = c_a/h_c$
            (h_s = sample height, h_c = height of diffusion chamber)
       K = slope of a plot relating the logarithm or relative concentration
            gradient to time

Neglecting the storage of diffusing gas in the sample during diffusion gives the simpler equation

$$D = - h_s \cdot h_c \cdot K$$

requiring knowledge of only the dimensions of the measuring device.

Time:
Measurements usually take 1—2 hours, depending on soil water content and soil structure. Frequency of data reading depends on required accuracy, initial and final values the absolute minimum. Calcuations relatively simple but computer desirable.

Costs:
Depending on gas measuring technique. If several sensors are required and technical expertise is not at hand the production of equipment can be expensive.

Accuracy:
Assuming reliable gas sensors and well controlled temperature the accuracy is reasonably good. No subjective aspects in the calculations.

Advantages:
Easy installation of soil sample in position for measurement (compared to methods 5.5.3. A—B).

Disadvantages:
Changing atmospheric pressure during an experiment can cause non-diffusive flow of gas through the sample. Evaporation from soil surface. Requires laboratory with well controlled temperature.

Remarks:
Use of oxygen as diffusing gas can cause erroneous results in moist respiring topsoils.

A variant of the method employs sampling of sub-samples of gas from diffusion chamber, reducing the need for many sensors.

References:
Evans, D. D. (1965). Gas movement. *In:* Black, C. A. (Ed.). Methods of Soil Analysis. ASA Monograph No. 9, 325-330.
Currie, J. A. (1960). Gaseous diffusion in porous media. Part 1.—A non-steady state method. British Journal of Applied Physics, 11, 314-317.
Bakker, J. W. and A. P. Hidding (1970). The influence of soil structure and air content on gas diffusion in soils. Neth. J. agric. Sci. 18, 37-48.

5.5.3 Non-steady state method: closed system

A. One chamber

Principle:
The sample to be examined is placed in one end of a closed chamber, the non-filled volume of which constitutes the diffusion chamber. A gradient of gas concentration is established between sample and chamber. Diffusion is then followed by measuring concentration in the chamber.

Equipment needed:
Sampling cylinders, field sampling equipment. (Equipment for potential-controlled drainage of samples). Sample holder/diffusion chamber. Radioactive gas ($^{14}CO_2$ used in the reference). Geiger-Müller tube. Computer.

Procedure:
Initially the system is filled with non-radioactive CO_2 and then a small amount of ^{14}C-labeled CO_2 is allocated to the diffusion chamber. Allowing exchange of gas between sample and diffusion chamber, the diffusion is followed by analyzing count rate of a Geiger-Müller tube.

Calculation.
From Fick's first and second laws an expression of the relation between the fractional amount of gas, which has diffused out of the sample (this amount is measured indirectly) and the ratio of D_{soil}/D_{air} can be obtained. The reader is referred to the reference for detailed explanation of calculation procedure, which involves reading of graphs etc.

Time:
Measurements take little time. Calculations time-consuming.

Costs:
Materials expensive because of the equipment for tracing the radioactive isotope. Highly qualified personnel needed.

Accuracy:
In principle good (see advantages).

Advantages:
Complete control of temperature and overall pressure. Little or no evaporation from the sample. No mass flow can arise, due to identical materials diffusing.

Disadvantages:
Complicated calculations.

Remarks:
Use of CO_2 as the diffusing gas can cause errors due to microorganisms evolving CO_2 as well as dissolution in soil water.

Reference:
Rust, R. H., A. Klute and J. E. Gieseking (1957). Diffusion-porosity measurements using a non-steady state system. Soil. Science, 84, 452-464.

B. Two chambers

Principle:
The soil sample is placed in the middle of a closed cylinder, with equal volumes of free space on both sides. Diffusion is achieved by establishing a concentration gradient and followed by measuring the concentration in both chambers at time intervals.

Equipment needed:
Sampling cylinders, field sampling equipment. (Equipment for potential-controlled drainage of samples). Diffusion apparatus. Gas measuring device (often photomultipliers in connection with radioactive isotopes).

Procedure:
A gas with a radioactive isotope, for instance Krypton-85 is injected into one of the chambers. Diffusion through the sample is followed by measuring the count of ß-radiation by photomultipliers in both chambers.

Calculation:
Fick's first law can also be applied to this system. With valid assumptions of boundary conditions a solution to the equation of Fick is:

$$C_I - C_R = 2C_e \exp \left(- \frac{2DA}{VL} \cdot t \right)$$

where C_I, C_R = concentration of the diffusing gas in the two chambers
C_e = concentration in the system when equilibrium has been reached
D = apparent diffusion coefficient
A = cross sectional area of sample
L = length of sample
V = volume of chambers
t = time

C_I and C_R are measured. By non-linear regression of $(C_I - C_R)$ and t, C_e and $\frac{2DA}{VL}$ can be estimated. Then also D can be calculated.

Time:
Measurements take 1—2 hours, depending on soil water content and soil structure. Frequency of data reading has to be rather high, but data acquisition can be automated.

69

Costs:
High. Need for photomultipliers (reference 3: gas chromatography). Computer needed. Highly qualified personnel needed.

Accuracy:
Good (see advantages).

Advantages:
Complete control of temperature and overall pressure. Little or no evaporation from the sample.

Disadvantages:
Expensive.

Remarks:
In one variant of the method (Weller et al., see reference) identical gases are used in both chambers, and one is marked with a radioactive isotope. This secures no mass flow in the system, due to identical diffusion coefficients of counter diffusing molecules.

Another variant make use of subsamples of air (analyzed by gas chromatography) from the diffusion chambers to determine the change in concentration (Reible & Shair, see reference).

References:
Weller, K. R. N. S. Stenhouse and H. Watts (1974). Diffusion of gases in porous solids. I. Theoretical Background and Experimental method. Canadian Journal of Chemistry, 52, 2684-2689.
Ball, B. C., the late W. Harris and J. R. Burford (1981): A laboratory method to measure gas diffusion and flow in soil and other porous materials. Journal of Soil Science, 32,323-333.
Reible, D. D. and F. H. Shair (1982): A technique for the measurement of gaseous diffusion in porous media. Journal of Soil Science, 33, 165-174.

5.6 Air permeability

5.6.1 Non-steady state method

Principle:
Air is compressed in a tank and allowed to flow through soil sample or down into soil. The decrease of pressure in the tank is measured as a function of time.

Equipment needed:
Sealed tank with air outlet tube to sample holder/cylinder for in situ measurements. Pump. Manometer.

Procedure:
See principle.

Calculations:
Combining the law of Darcy for laminar flow and the law of Boyle/Mariotte for pressure/volume-relation, with certain assumptions, yields:

$$\ln\Delta P = -K \cdot \frac{A \cdot P}{\eta \cdot L \cdot V} \cdot t + \text{Constant} \qquad (1)$$

where K   = permeability cm^2
      A   = cross sectional area of sample cm^2
      L   = length of sample cm,
      P   = atmospheric pressure $N\ m^{-2}$
      η = viscosity of air $N\ .\ s\ m^{-2}$
      V   = volume of air tank cm^3
      t   = time s,
      ΔP = pressure difference between tank and atmosphere, arbitrary dimension.

For in situ measurements an empirically determined constant is used instead of the ratio A/L. Registrating ΔP and t, K can be calculated from (1).

Time:
Measurements take only little time, but several data readings are necessary.

Costs:
Material costs are low. Highly qualified personnel not needed.

Accuracy:
In comparison to method (5.6.2), several steps in measurement and calculation procedure can cause error.

Advantages:
Robust equipment for field measurements.

Disadvantages:
The necessary pressures used can cause some of the water films blocking pores to rupture and also create turbulent flow.

Reference:
Kirkham, D. (1947): Field Method for Determination of Air Permeability of Soil in its Undisturbed State. Proceedings of the Soil Science Society of America, 11, 93-99.

5.6.2 **Steady state method**

Principle:
Air is forced through a soil sample or down into soil at a constant pressure by the movement, due to gravity, of a floating cylinder. Flow is calculated as volume of air displaced in a time unit.

Equipment needed:
Apparatus with a floating cylinder, allowing controlled and measurable flow of air to a sample holder/cylinder for in situ measurements.

Procedure:
See principle.

Calculations:
The law of Darcy can be applied directly:

$$Q = K \cdot \frac{P\ .\ A}{\eta\ .\ L} \cdot t, \tag{1}$$

where   Q   = volume of air displaced at time t cm^3
          K   = permeability cm^2
          P   = applied pressure difference $N\ m^{-2}$
          A   = cross sectional area of sample cm^2
          L   = length of sample cm,
          η = viscosity of air $N\ .\ s\ .\ m^{-2}$

For in situ measurements an empirically determined constant is used instead of the ratio A/L. K can be calculated from equation (1), when a value of Q at time t has been determined.

Time:
Measurements take little time, only one data-reading necessary.

Costs:
Material costs are low. Highly qualified personnel not needed.

Accuracy:
Good. Few sources of error.

Advantages:
Only small pressure differences necessary. Measurement requires only a flow determination, which is reduced to reading a stopwatch. K follows directly from this value.

Disadvantages:
Rather vulnerable equipment for field trip (compared to method 5.6.1). In the standard method with a floating cylinder, the driving pressure decreases slightly during a measurement, due to increasing buoyancy. This has not been taken into account in calculations.

Remarks:
Variants of the method are controlling the air flow and pressure by precision pneumatic controller and controlled flow of water, respectively (Ball et al., 1981, Andersson, 1969).

References:

Grover, B. L. (1955): Simplified Air Permeameters for Soil in Place. Proceedings of the Soil Science Society of America, 19, 414-418.

Ball, B. C., the late W. Harris and J. R. Burford (1981): A laboratory method to measure gas diffusion and flow in soil and other porous materials. Journal of Soil Science, 32, 323-333.

Andersson, S. (1969): Markfysikaliska undersökningar i odlad jord, XIX. Grundförbättring, 4, 143-154.

CHAPTER 6

Soil strength and stability

6.1 Introduction

A number of properties that define the physical state of the soil have a semi-permanent character: the general structure of the profile, granulometric constitution, mineralogy and—to a lesser extent—organic and ionic constituents of the different horizons. Interacting with those quasi stable properties, a second group of state characteristics are susceptible to a rapid and sometimes very important change under the influence of external climatic and cultural factors that are particularly active in the most exposed surface layers.

Soils react to those factors in different ways, and for several decades soil physicists have concerned themselves with explaining these specific behaviours and have proposed methods for evaluating them, or at least for classifying them.

The principal difficulty arises because the climatic and cultural actions on the soil are complex and result from elementary factors (variation in soil water potential, the level of mechanical strains . . .), the real modes of application of which to the soil are poorly understood.

Confronted with this complexity and with our actual inability to take it into account, we are lead to regroup these actions about the three principal groups of processes which involve:
—The actions of fragmentation (fissuring by swelling and shrinkage, breaking up by machines for working the soil . . .) accompanied or not by sorting of the elements that arise.
—The actions of compaction (rolling, treading by animals, pressure exerted by working tools, phenomena of differential swelling and shrinkage).
—The actions of disaggregation by water of the structural units, which can go as far as the dispersion of the colloidal minerals and organic constituents.

Meaning and limitations of criteria for behaviour

Three principal groups of properties (Monnier et al, 1982) influence the behaviour of soils in relation to these actions:
—The structural stability, a property which affects the phenomenon of disaggregation by water.
—The property of swelling-shrinking influencing the processes of fissuring.
—The behaviour in relation to mechanical actions and especially in relation to compaction.

At this point also another difficulty arises: considerable differences are sometimes recorded between observed behaviour in situ and the results of laboratory tests. These differences can result from diverse causes:
—on one hand the means of applying these actions during laboratory tests differ quantitatively and qualitatively from those that prevail in the field.
—on the other hand, the initial conditions (the structural and moisture status during the application of the action) play, in combination with the intrinsic properties of the material, a role that is occasionally predominant.

Criteria for the choice of reference methods

These difficulties have brought about, on the part of the laboratories working in this domain, two types of response that are paradoxically opposed.

A. In the case of structural stability there exists a great variation of methods: (26 methods reported in "West European Methods for Soil Structure Determination, 1967"). The analysis of this variation in methods made evident that the differences depend principally on:

(a) **The sample which is subjected to the test**
This, it seems, is one of the most fundamental differences which gives rise to two major types of concept:
   —on the one hand, the sample subjected to the test is in the state in which it existed on the ground at the moment at which it was taken. Such tests take into account simultaneously the influence of the size of the clods and the intrinsic stability of the constituent material. Their results are therefore at least partially dependent on the particular condition of the soil at the time of sampling.
   —on the other hand the sample taken is previously altered to a standard structural state. This concept favours considerably the permanent properties or those which change very slowly (texture, organic matter content, ionic composition).
The objectives of interpretation and the possibilities of utilizing the results are at this point different for the two types of test: predicting a short term structural evaluation in the first case, and the evaluation of the degree of risk and comparison of properties in the second—so that we must retain two reference methods corresponding to one or other of these concepts (Henin, 1958, De Leehneer, 1959).

(b) **The treatment to which the sample is submitted during the test**
This is generally rigidly standardised but its nature is very variable: wet sieving with or without previous treatment and agitation; utilization of artificial rainfall; utilization of mild dispersants, . . . etc.
   If one recognises well the usefulness of one standardised treatment (reproducibility of results and validity of comparisons) one must emphasise that it does not account for the diversity of ways that water acts "in situ".
   Also we have retained, for one of the references, a method that proposes by means of pretreatments before wetting, a scale of severity of treatment by water in relation to the disaggregation.

B. Opposed to this diversity of methodologies, the mechanical properties and the ability of earthy materials to change volume with their water content have not achieved, until very recently, the objective of a small number of proposals. The choice of one reference method depends then, less on the criteria for the representation of different objectives and the mean of achieving them, than on the intrinsic qualities of the methods available.

(a) In that which concerns the Mechanical Behaviour, research workers often have recourse to the standardized determination of "Consistency Limits" (Atterberg limits). These tests are simple and not expensive but one must only consider their results as relative indicators of mechanical behaviour that complement the comparisons of granulometric constitution.
   In the more specific domain of behaviour towards compaction, one may have recourse to two principal types of tests:
   —the tests of dynamic compaction, of which the Procter test is the best known, have the advantage of being easy and cheap. By contrast, it is not adaptable for the compaction of undisturbed samples; these produce irregular zones of compaction that are systematically heterogenous and do not easily permit selective variations of levels of energy to be applied. Finally and most important, they do not permit the study of the kinetics of compaction nor those of the phenomena of recovery, in particular of the important role of the soil organic matter in this regard. In spite of these limitations, we propose the Proctor test as a reference method on account of its practical advantages.
   —the tests of static compressibility (oedometry) do not present any of these disadvantages: moreover they allow work on consolidation, and provide the means adopted for morphological or morphometrical analysis. It is for this reason that despite their very high cost (special apparatus linkable to data acquisition systems), it is possible to consider oedometry as a good reference method.

74

(b) In the case of measurements permitting judgement of the capacity of swelling and shrinking, the methods depend totally on the establishment of all or part of the curve defining the relationships between the bulk density of a sample and its water retention.

The variations depend mainly on:
—the nature of the sample on which the measurements are made: initially saturated paste in the case of shrinkage curves;
—undisturbed aggregates or clods.
—the methods of measuring bulk density at different water contents.
—the methods of getting the measurements of water retention and the direction of imposed moisture variations.

The method (Stengel, 1982) proposed as a reference depends on natural aggregates of small size. The water retention values, obtained by equilibrating potential, extend from zero to moisture contents corresponding to very weak potentials.

Applications and methods of using results for specific objectives

It follows from the preceding that the physical behaviour of a soil in situ during a determined time interval can in theory be described by a function, the variables of which reflect the actions being exercised on the system during the period under consideration (rainfall characteristic, level of mechanical strain, variation of water potential . . .) and parameters which are representative of the specific properties of the material (intrinsic behaviour) and its initial physical state.

Depending on the level of information, the practical possibility of measuring the variables and the parameters and the type of phenomena to be studied, one can envisage such models in a deterministic or a more or less complex statistical form.

We propose here to make some brief remarks on some problems arising from three major types of behaviour, which are of major importance in the actual context of member countries of the EEC, the tilling of the soil, the erodibility of cultivated soil by water, the trafficability of agricultural lands.

(a) With regard to that which concerns the working of the soil one of the principal questions concerns the suitability of the soil for sequential direct drilling. In this area the approach might consist of developing a method to predict the evolution of the mean physical state of soils in the absence of all mechanical intervention agents.

If at the first stage one neglects the influence of the action on the soil of variations in climate and cultivations, one can propose (Stengel et al, 1984) by a combination of the levels of 3 principal behaviours: (structure stability, behaviour towards compaction, liability to fissuring), a classification of the lands in relation to their suitability for direct seeding confirmed by cultivation results. The limit of such an approach is seen clearly: it only gives mean information, the validity of which can be called into question when in any one year the climatic conditions depart sufficiently from normal conditions but also when the soil conditions not taken into account by the 3 criteria retained (quality of drainage for example), affect with significant influence the physical or agronomic behaviour.

(b) A second group of questions is posed by the superficial structural degradation by the rain and its effects indeed on the functioning of seed beds and on the loosening by streaming and water erosion on sloping ground. In either case one rediscovers the dialectic relationship between the climatic variables and the behavioural parameters of the evolution of the soil. It is thus that one speaks of pelting rain (having high Kinetic energy) and of beaten soils (having weak structural stability); similarly in that which concerns water erosion, its prediction brings in the concept of erosivity of rain (active variable) and that of erodibility of soils (parameters of susceptibility).

The central point of the two processes is the evolution of the infiltration flux affected by the evolution of the surface (appearance of structural crusts and, on flat ground deposition). It is clear that acting as a control of seed beds, the structural evolution is the determining factor on the agronomic effectiveness of seed bed preparation techniques: the germination and the emergence of seedlings are closely related to the properties for transfer of water, gas and heat and the mechanical properties of crusts that are likely to form. The parameters of the initial state and of the intrinsic structural stability of the materials takes into account the importance of all of them.

In the case of research on erosion the variable of entry is more often than not run off. Then, the idea of erodobility and its principal components of detachability and transportability has a tendency to disappear in the presence of ideas of the erosivity of rain and (or) of run off. The phenomena of erosion which has recently appeared on loamy soils on gentle slopes in temperate humid climates, if correlated with the reduction in the content of organic matter and the structural stability, suggests more or less that one must not neglect the erodibility characteristics of the material.

(c) The last example which we meet is that of the research of diagnostic criteria and by means of this to predict the trafficability of cultivated ground. This term implies the taking into account of theories of mechanical nature (bearing capacity, adhesion) and the agronomic concern of physical nature (effects of packing, inevitably caused by wheeling, on the functioning of the soil and its cultural performance). The observation and the analysis of the deformations of soils caused by wheeling clearly shows that the conditions satisfying these two types of concern are not always compatible.

This fact arises essentially from the existence of phenomena of lateral flow and of the role played by "layered" beds with low deformability. The difficulty of predicting trafficability is further increased in the presence of strong gradients of water content in the superficial soil layers particularly in Spring.

The results of oedometric tests are able (Guerif, 1982) to provide comparative references on the behaviour under load (or after relaxation) of materials, as a function of their state of hydration and their structural state, of the pressure exerted and its duration of application.

Finally, the physical evaluation and more especially the agronomy of packing, does not proceed from one instantaneous bit of information; the judgement of trafficability taken in relation to risks of packing, must also take into account the possibilities of regeneration of which we have met certain aspects appropriate to the properties of swelling and shrinking of the soil.

In short, the evaluation of the principal mechanical and structural behaviour of the soil by the measurements of their reactions by normal treatments may lead to a range of stages of utilization between two poles of methodologies.

A first approach accepts a certain statistical stability of climatic and cultural conditions. Levels of behaviour which should be evaluated, are considered, either in terms of comparisons between different soils, or as indicators of the evolution of the properties of a particular soil over a period of time.

A second approach uses descriptive models to predict the behaviour of soils in situ by introducing the real forces applied to the soil. In models of this kind the influence of many available elements (Boiffin, 1984) may already have been quantified statistically in terms of intrinsic behaviour.

It goes without saying that between these two poles combined solutions can apply according to the state of knowledge of the concrete response expected from the model.

In any case, it appears to be immediately evident that there is an interest in selecting reference methods which enable research workers to compare the results of large numbers of methods without inhibiting the further advances in methodology which are still essential.

6.2 Literature cited

Monnier, G., P. Stengel, and J. Guerif, (1982). Recherches de critères de la fertilité physique du sol et de son évolution en fonction du système de culture. C.R. Symp. Fertility of Soils, C.E.E., BARI, P. 35-52.

Henin, S., G. Monnier, and A. Combeau, (1958). Méthode pour l'étude de la stabilité structurale des sols. Ann. agron. 9(1), 73-92.

De Leehneer, L., and M. De Boodt, (1959). Determination of aggregate stability by the change of the mean weight diameter. Proc. of the Inter. Symposium on Soil Structure, Ghent, p. 290-300.

Stengel, P. (1982). Swelling potential of soil as a criterium of permanent direct drilling suitability. 9e Conf. ISTRO, OSIJEK (Yug.), 131-136.

Stengel, P., J. T. Douglas, J. Guerif, M. J. Goss, G. Monnier, and R. Q. Cannel (1984). Factors influencing some properties of soil in relation to their suitability for direct drilling. Soil and Tillage Research (4), 35-53.

Guerif, J. (1982). Compactage d'un massif d'agrégats. Effet de la teneur en eau et de la pression appliquée. Agronomie, 2(2), 287-294.

Boiffin, J, (1984). La dégradation de la surface du sol sous l'action des pluies: Etude d'un comportement du sol au champ. Thèse, Paris I.N.A.-P.G.

6.3 Structural stability

6.3.1 Wet sieving

Objective:
To make a comparison of the behaviour of the constituent materials of soils in relation to disaggregation and dispersion by water.

A. Water-alcohol-benzene treatment

Procedure:
200 g of air dried soil are hand sieved through a sieve with 2 mm square holes. Three samples of 5 g are taken and put in separate beakers or erlenmeyers and either 10 ml of ethanol or 10 ml of benzene added or a large quantity of distilled water. After the suspensions have settled for 30 minutes, water is added to a volume of 300 ml and the samples are shaken end over end 20 times. After agitation and standard sieving the amount of "stable" aggregates remaining on the 200 µm sieve is determined for the different treatments and also the amount of clay and silt (fraction <20 µm) that remain in suspension.

Calculations:
The instability index Is (or S) is obtained as follows:

$$\text{Instability Index} = \frac{(\% < 20\,\mu m)_{max}}{\dfrac{Ag_a + Ag_b + Ag_c}{3} - 0.9\ (\%\ \text{course sand})}$$

where: Ag_a = % stable aggregates after the alcohol test
  Ag_b = % stable aggregates after the benzene test
  Ag_c = % stable aggregates after the water test (untreated)

The S index might vary from 0.1 for the most stable soils (rich in organic material) to values >100 for very unstable soils (sodic).

Equipment:
Standard wet sieving apparatus.
Sieves of 2 mm and 200 µm.

Remarks:
The maximum quantity of particles <20 µm is obtained in the suspension treated with benzene or in the suspension directly treated with water. The method is not suitable for predicting short term evolutions of systems affected by rainfall impact and initial moisture content.

Disadvantages:
The use of benzene can cause some health problems. Long term storage of samples rich in organic matter can result in development of by-products.

References:
Hénin, S., G. Monnier, and A. Combeau, (1958). Méthode pour l'étude de la stabilité structurale des sols. Annales agronomiques, 1, 73-92.
Féodoroff, A. (1960). Evaluation de la stabilité structurale d'un sol (indice S). Nouvelles normes d'emploi pour l'appareil à tamiser. Annales Agronomiques 11, 651-659.

77

B. Water treatment

Procedure:
Air dried soil clods are sieved between 8 mm and 2 mm for clay and loam soils, and between 8 mm and 1 mm for sandy soils and the dry aggregate size distribution determined. Each aggregate size group is wetted under simulated single drop rainfall. After equilibrium the moistened aggregates are sieved under water. After sieving, each aggregate size group is dried, weighed, and the aggregate size distribution determined.

Calculations:
The instability index is determined as the measured surface area between the curves corresponding to the aggregate size distribution before and after sieving. This is numerically the change in mean weight diameter between the dry aggregate distribution and the wet stable aggregate size distribution. The larger the value the more unstable are the aggregates.

Equipment:
Standard dry and wet sieving apparatus with sieves between 8 mm and 1 mm.

Remarks:
This method takes partly into account the effect of detachment by raindrop impact and is suitable for studying seedbed preparation.

References:
De Leenheer, L., and M. De Boodt, (1959). Determination of aggregate stability by the change in the mean weight diameter. Proceedings of the International Symposium on Soil Structure, Ghent 1958, Mededelingen van de Landbouwhogeschool, 24, 290-300.

De Boodt, M., L. De Leenheer, and D. Kirkham, (1961). Soil aggregate stability indexes and crop yields. Soil Science, 91, 138-146.

6.4 Swelling and shrinking of aggregates

Objective:
Evaluation, from the amplitude of swelling-shrinking, of the liability of soils to fragmentation by fissuring.

Principle:
Water content and bulk density of aggregates are measured after wetting and at different stages of drying.

Procedure:
Aggregates are wetted at low pF (pF 0.48: potential wetting state). The bulk density of aggregates at different stages of drying is determined by densiometry (cf. reference method: 4.2.4. Bulk density of aggregates).

Calculations:
The moisture-bulk density curve is determined and the swelling-shrinkage index is calculated as:

$$\frac{\text{bulk density at pF 0.48} - \text{bulk density at shrinkage limit}}{\text{bulk density at shrinkage limit}}$$

Remarks:
The complete shrinkage curve permits the evaluation of shrinking-swelling irrespective of the range of moisture content. It also permits the prediction of development of porosity by cracking as a

function of moisture content in soil layers at constant density. The total volume of cracks can be determined during the different stages of swelling and shrinking. Neither the dimensions and network of fissures nor the anisotrophy of swelling and shrinking can be determined.

Variation to the method:
Kuipers, H. (1961). Preliminary remarks on porosity of soil aggregates in an air-dry state and at pF 2. Neth. J. Agric. Sci. 9, 168-173.

Reference:
Stengel, P., (1982). Swelling potential of soil as a criterium of permanent direct drilling suitability. C.R. 9th Conf. of ISTRO, OSIJEK, 131—136.

6.5 Compaction tests

6.5.1 Oedometric test

Objective:
To evaluate the effect of moisture, applied pressure and mineral and organic constituents on compaction of different soils.

Principle:
An uniaxial laterally confined pressure is applied to disturbed or undisturbed soil samples. The intensity and duration of the applied pressure can be changed, and the change in sample volume measured under load (compressibility) as well as without load (recovery). If necessary, water can be evacuated during the compaction test (consolidation).

Procedure:
A soil sample placed in a cylindrical cell is compressed by a piston pneumatically controlled or moved by the action of weights. The change in sample volume can be recorded as a function of time.

Calculations:
The change in bulk density or in pore volume can be determined as a function of different factors.

Equipment:
Oedometer cell.
Compression apparatus.
Recording system.

Duration of test:
The total duration of a test is 24 hours per sample but the effective working time is considerably less.

Advantages:
Possibility of tests with drained soil.
Possibility of unique variations in pressure.
Homogeneity and thus the possibility of using test tube samples.
Possibility of using undisturbed and small samples.
Possibility of changing the initial state of the sample.

Disadvantages:
Costly and very time consuming.

Areas of application:
The behaviour of soils under compaction in saturated and unsaturated conditions (with consolidation).
Determination of consistency limits (in analogy with Atterberg limits) according to different energy levels.

Reference:
Guerif, J., (1982). Comportement au compactage d'un massif d'agrégats. Effet de la teneur en eau et de la pression appliquée. Agronomie, 2(2), 287—294.

6.5.2 Proctor test

Objective:
To evaluate the consistency limits to a standard energy and the effect of organic and mineral components on the mechanical behaviour of soils.

Principle:
Soil samples packed in a mold at different moisture contents are subjected to standard dynamic compaction by means of a falling weight. The bulk density of the compacted material is determined.

Procedure:
The dynamic compaction test is carried out on different soil layers. If a mass m falls N times from a height H on each of the n layers, the applied energy is:

$$E_c = \frac{m.g.H.n.N}{\text{volume of mold}}$$

For agricultural purposes the mold has a volume of 300 cm^3 and an inner diameter of 70 mm and m = 1 kg, H = 20 cm, n = 3 and N = 20.
The standard applied energy is 590 kJm^{-3} or 60 T/mm^{-3}. The bulk density of the soil sample at different moisture contents after compaction is determined.

Equipment:
Proctor mold and weight.

Variation:
Standard Proctor mold for Civil Engineering Tests is 947 cm^3.

Advantages:
Relatively cheap compared to oedometric test.

Reference:
Faure, A. (1978). Comportement des sols au compactage; Rôle de l'argile et conséquences sur l'arrangement des grains. Thèse Université Scientifique et Médicale de Grenoble.

6.6 Measurement of consistency limits (Atterberg limits)

Objective:
To determine physical constants of soils in terms of workability as a function of soil moisture content.

Principles and definitions:

The liquid limit is defined as the water content at which a V-shaped grove cut in moist soil held in a special cup of the Cassagrande apparatus is closed after 25 blows. The plastic limit is the water content at which the soil begins to crumble when rolled between the hand and a glass plate into a thread of about 5 mm. It represents the lowest water content at which soil can be deformed readily without cracking. The plasticity index is the amount of water which must be added to change a soil from its plastic limit to its liquid limit and is an indication of the plasticity of the soil.

Equipment:

Apparatus of Cassagrande, glass plate.

Duration of test:

For a skilled operator, the time required for a single determination of the liquid limit by the method of Cassagrande is about 30 minutes. As this has to be done repeatedly before the corresponding moisture content at which the grove is closed by 25 blows can be determined, a one point method has been proposed.

Remarks:

The state of energy put to work during the test is weak compared to agricultural rolling for example. The state of energy is also different for both limits (plastic and liquid) in the plastic range. On the contrary, the moisture-density curves drawn from Proctor tests and oedometer measurements can be used to evaluate the two limits of consistency at the same level of energy.

References:

Lambe, T. W. (1967). Soil testing for Engineers. John Wiley & Sons, Inc.

Sowers, G. F. (1965). Consistency. In: Methods of Soil Analysis. Ed. C. A. Black, Agronomy No. 9, A.S.A., p. 391—399.

Mohan, D. and R. K. Goël, (1961). One point method of determining liquid limit. Soil Science 91, 100—102.

6.7 Penetration resistance

Objective:

To evaluate soil working operations and soil compression.

Principle:

The resistance offered by the soil to the penetration by a metal probe is recorded continuously or discontinuously. The probe is usually placed on the soil surface and moved downward through the profile.

Procedure:

A cone-shaped probe is driven into the soil either by a permanent force or through consecutive falls of a given load (hammer sliding along a rod). The resistance is recorded as the force necessary for the probe to penetrate to a certain depth, or the depth of penetration under a given load, or after a certain number of falls is registered.

Equipment:

A variety of equipment is available. Most commonly used is a cone with a 60° angle and a cross section of 1 cm^2 to 5 cm^2 for very soft soils. Mechanisms of force transmission vary from a hand-operated hammer over computarized systems to machine driven rams.

Calculation:

Results are frequently expressed as pressure (bar, Pa) if the equipment allows a force to be measured. A penetrometer with hammering or ramming action only allows the registration of the number of falls or strokes or penetration depth.

Costs:
Basic costs for the most simple equipment are low. The more sophisticated the registration device, the more expensive the equipment.

Accuracy:
Because of the inherent properties of this method the heterogeneity of the soil is far greater than any possible inadequacy of equipment. So if deviations from means are calculated the results reflect soil properties more clearly than apparatus shortcomings. Variation of penetration resistance is frequently geometric in scale rather than normal (arithmetic).

Advantages/Disadvantages:
Measurement of penetration resistance is simple. The equipment is generally easily carried over the area in question. There is no other physical method to cover areas from several square meters up to several hectares so rapidly to obtain a preliminary view of the general situation in depth as well as in a horizontal direction.

This advantage allows the method to be looked at as a preliminary tool used for example to find the exact locations for more tedious sampling of in-situ measuring methods. (Hartge, Bohne, 1985).

By using penetration resistance it is easy to find average or extreme situations for further investigation. This method should be envisaged in the same way as seismic or geoelectric methods when tracing changes in rock quality in oil-exploration.

Disadvantageous is the fact, that penetration resistance of a soil depends on many properties. Therefore it is extremely seldom that it can be used to single out a particular one and in most cases it is not useful to calculate results to give physically basic dimensions in place of numbers of blows or depth of penetration. Physical calculations tempt to overinterpretation and consequently to disappointment.

Over-interpretation in the past is the most frequent reason for low interest in this method, the merits of which lie in its speed and simplicity.

Remarks:
Penetration resistance is not a particular property of a soil, but rather a summation effect of several properties of which bulk density, water content and shearing resistance are the most important ones. These basic soil properties are, at the moment of penetration measurement, not only dependant on structure but also on grain size distribution, mineral composition, water regime history (drying or wetting) and amount and character of organic matter.

This complex situation makes it unlikely that results could be meaningful in any other direction than: "Here is a difference at this location—investigate its nature with other methods". But in this respect the method is unique.

References:

Kullmann, A. (ed.), (1968). Untersuchungsmethoden des Bodenstrukturzustandes VEB. Deutscher Landwirtschaftsverlag (Inter. Soc. Soil Sci. Comm. I,East European Group).

De Boodt, M. (Ed.), (1967). West European Methods for soil structure determinations (Intern. Soc. Soil Sci., Comm. I West European Group).

Hartge, K. H., and H. Bohne, (1985). Penetrometer measurements for screening soil physical variability Soil Tillage Res., 5,343—350.

ESOPT (1974). Proceedings of theEuropean Symposium on Penetration Testing in Stockholm 5-7/6/74 2 Vol statens Råd fur Byggnadsforskning S-11460 Stockholm.

CHAPTER 7

Soil morphology

7.1 Introduction
Soil morphology studies can make original contributions to the characterization of soil structure. Soil profile descriptions, as reviewed in chapter 1, include a field evaluation of soil structure by noting sizes, shapes and degrees of development of natural aggregates (peds). Also, the occurrence of macropores formed by roots or soil animals is observed. This information is valuable, in principle, to improve soil sampling procedures for physical structure measurements, as discussed in chapter 2. However, soil morphological studies can also contribute more specific, quantitative information by analysing soil thin sections or polished sections of plastic-impregnated, undisturbed soil samples. Recently, submicroscopic techniques have been developed that allow very detailed observations and chemical and mineralogical characterizations of soil compounds, as occurring in situ. Such techniques may be valuable for investigating structure stability.

Of particular significance is the application of morphological techniques in ways yielding data that could not have been obtained by physical methods. For example, the study of different types of pores, shapes of peds and large-pore continuity can only be made by morphological techniques. The efficiency of such studies is increased when they are made together with soil physical studies, which always in the end have to yield the type of data that are needed to characterize the occurrence and flow of water and air in soils.

The following subchapters will focus on the following aspects:

Preparation techniques
7.2 Preparing soil lacquer peels and soil monoliths
7.3 Preparing soil thin sections

Analysis of thin sections
7.4 Morphometric characterization of thin sections or polished sections of soils
7.5 Submicroscopy

Functional characterization of soil pores
7.6 Macropore-continuity
7.7 Use of soil morphology to calculate K_{sat}.

7.2 Preparing soil lacquer peels and soil monoliths
Principle for lacquer peels: A thin film of diluted lacquer is applied by a spray or brush onto the smoothed face of a profile pit. After one or more layers of lacquer have been applied and allowed to dry, the film, usually reinforced by cloth, is loosened from the face. The soil material will remain attached to the hardened lacquer, thus giving a natural replica of the soil profile in the form of a thin lacquer peel or lacquer profile.
Equipment needed: Spades, knives, profile-lacquer.
Procedure: The reader is referred to a very detailed step-by-step outline by Van Baren and Bomer (1979).

Time required: Limited when soil conditions are favourable, and when the method is applied in the context of a field survey.

Cost: Limited.

Advantages: Method yields a rapid, permanent picture of the natural structure in sandy or light clayey soils, that were dry at the time of sampling. Peels are light in weight and easy to handle, exhibit and store when mounted on a board. Soil structure can be studied with binocular-microscopes on soil lacquer peels.

Disadvantages: Soils with peds cannot be preserved this way because peds are not impregnated. The method cannot be used in very moist or wet soil. The lacquer needs at least half a day to dry. It is therefore necessary to return the next day to collect the lacquer peel.

Remarks: The soil monolith method can be used to overcome problems mentioned above. This method has the following steps:

The profile is collected from the pit in a box and taken to the laboratory or workshop for preparation and preservation. The soil is left to dry before the lacquer is applied, using a similar procedure to that for lacquer peels, but in controlled conditions. The depth of the impregnation is greater because the box with the soil is placed horizontally and several coatings of lacquer are applied at intervals.

The resulting soil monolith usually has a thickness of several centimeters, showing, in almost all cases, the structure of the soil very clearly. This method is suitable for all soils, except very sandy ones, because the almost vertical face of the soil profile may collapse during the field collection. The soils may be collected at any time, even when it is raining or when the moisture content of the soil is high. Collection of a soil monolith can usually be done in a single day and no chemicals are needed in the field. A disadvantage is the heavy weight of the box with soil material (30-60 kg), and the possibility of expensive transportation.

References

Jager, A. and van der Voort, W. J. M., 1966. Collection and preservation of soil monoliths. Soil Survey Paper 2, Soil Survey Institute, Wageningen.

Van Baren, J. H. V. and W. Bomer. 1979. Procedures for the collection and preservation of soil profiles. Technical Paper 1. ISRIC, Wageningen, The Netherlands. 23 p.

7.3 Preparing soil thin sections

Principle: An undisturbed piece of soil is impregnated by a liquid plastic with added chemicals that induce hardening after a certain period of time. Ideally, all soil pores are filled with the hardened plastic, allowing cutting and grinding without disturbance of the original arrangement of the soil components.

Equipment needed: Expensive equipment is needed for impregnation, cutting and grinding, as is explained in the reference papers.

Procedure: The basic procedure for the preparation of thin sections consists of the following phases:
1. Drying of the samples
2. Impregnation and hardening
3. Sawing and grinding.

Ad 1 Air drying is the most simple method. However, unnatural shrinkage may then occur in clayey soils. To avoid this, **replacement of water in wet samples** may be more attractive.

In the last decennia a number of methods have been developed to replace soil moisture e.g. (Miedema et. al., 1974; Singh, 1969). The method elaborated by Miedema et. al. is often used. The wet samples are placed in a container and acetone is added until the sample is submerged. About once a week the acetone/water mixture is replaced by pure acetone until no water is detected in the liquid syphoned off.

Another method is **freeze-drying,** which consists of two steps: The freezing of the sample at low temperature and drying of the sample via evaporation of ice and sublimation of the ice vapour on an ice condensator. The common method is developed by Jongerius and Heintzberger (1975). The freezing of the samples (max. 2 cm thick) occurs at $-158°C$ and takes about 3 minutes, the drying about one week. From the surface of the sample thin sections are made.

Ad 2 The composition of the impregnation resin differs, depending on the nature of the material to be impregnated. Commonly, the following mixture is used to impregnate 4 mammoth sized samples (15 × 8 × 5 cm) in one box, often a stainless steel box:

2500 cc Synolith polyester resin no. 544 S 36
2500 cc Acetone (diluent)
   5 cc Cyclohexanonperoxide, also named cyclonox LNC (catalizer)
   3 cc Colbaltoctoate 1% (accelerator).

Other chemicals are used as well. The box with samples is placed into a vacuum disiccator or other impregnation device. After evacuation of air with the vacuum pump, the resin is added until the samples are submerged. After release of the vacuum the box is transferred to a well ventilated fume cupboard where the added diluent evaporates. The samples are left in the fume cupboards for about 4-6 weeks to harden.

Ad 3 The hardened sample is sawed into flat slabs. A grinding machine is used to reduce the thickness of the slab, which is mounted on an object glass, to approximately 50 um. With handgrinding a thickness of 20 um is reached. This thickness is needed for micromorphological studies using transmitted light. The thin section is ready for use after mounting of a coverglass on top of the sample or by polishing of the surface. A flat sawed face of a cut slab can also be polished.
Time required: As indicated above, the procedure is very time consuming.
Cost: High.

References

Jongerius, A. and G. Heintzberger, 1975. Methods in soil micromorphology. A technique for the preparation of large thin sections. Soil Survey Paper no. 10, Netherlands Soil Survey Institute, Wageningen, The Netherlands.

Miedema, R., Th. Pape and G. J. van der Waal, 1974. A method to impregnate wet soil samples, producing high quality thin sections. Neth. J. Agric. Aci. 22: 37-39.

Singh, R. B., 1969. A versatile method for treatment of clay soils for thin section fabric studies. J. Soil Sci. 20: 269-273.

7.4 Morphometric characterization of thin sections or polished sections of soils

Principle: Morphometric characterization of materials prepared into thin sections consists of two phases. First, a micromorphological study of thin sections is made by means of light microscopy or submicroscopy (see method 7.5). After identification of relevant features and their description according to standard procedures, quantification of morphological features is obtained. Quantitative information can be obtained most accurately by using an electro-optical image analyzer (e.g. Quantimet). Polished sections of soils can be studied in the same way.

Equipment needed: The basic study of thin sections is performed with a polarization light-microscope, using magnifications up to x200. Transmitted light as well as reflected (incident) light can be used for optical identification of morphological features. Direct quantification via light microscopy occurs by incorporation of different grids in the eyepiece (ocular) of the microscope that allow point or line counts. Sophisticated quantification is achieved by using electro-optical image-analyzers e.g. a Quantimet. Image analyses can be performed directly from thin sections or from special photographs.

Procedure: A detailed review of description schemes of soil thin sections is beyond the scope of this text. The reader is referred to a recent publication by the working group of the subcommittee on soil micromorphology of the International Soil Science Society (Bullock et. al. 1985). An abbreviated, more operational version was published by Kooistra (1985). Procedures will be illustrated for soil pores, which form a crucial part of soil structure. Figure 7.1 shows a schematic representation of types of soil pores. Occurrence of natural soil aggregates ("peds") is indicated by the term "pedality". Pedal soil materials contain peds, apedal soil materials don't. Soil pores that are larger than the simple packing pores of the individual soil particles may be **planes** ("cracks", which form by shrinkage of clayey soil materials) **channels** (cylindrical pores formed by soil animals or roots), **compound packing voids** (packing pores between rounded aggregates) or **vughs** (all other larger pores). All pores larger than 30 µm are referred to as macropores. Five types of microstructure with different types of micropores are distinguished as a function of the arrangement and packing of the

VOIDS IN SOILS

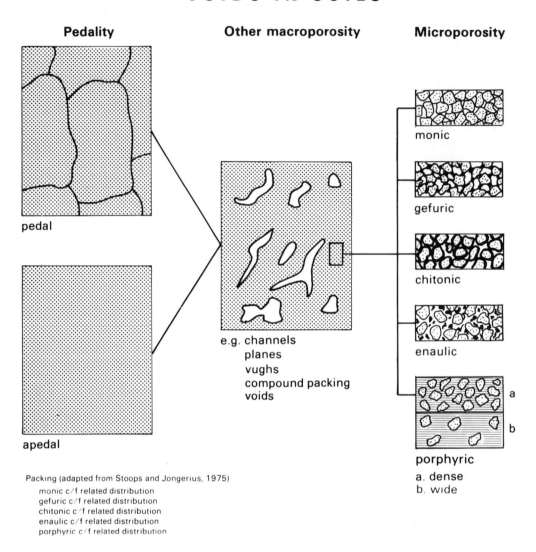

Fig. 7.1 Schematic presentation of types of voids present in soils.

basic soil constituents (Fig. 7.1). Quantification of soil macroporosity is illustrated in Fig. 7.2 which shows three thin-section images of a soil horizon between 30 and 40 cm depth in a sandy loam soil, under three different types of management: (A) Grassland, (B) Arable land, resulting in formation of a plowplan and (C) Arable land in which the plowpan was broken by deep tillage. Channels are found in soils A and B, and vughs in soil C, where the initial compound packing voids between tillage-fragments have collapsed forming discontinuous vughs. Fig. 7.3 shows results of a Quantimet measurement, indicating the percentage by volume of macropores of various size-classes. Quantimet measurements involve electronic scanning of a microscopic image, using the grey value of each of 600 000 picture points per image. Each grey value for a picture point corresponds with an electric pulse. A detector and a pattern recognition system are used to classify the pulses, allowing the distinction of two dimensional structural units that can be classified in terms of several parameters such as: area, perimeter, intercept etc. Further details are reported by Jongerius (1973); Ismail (1975); Bullock and Murphy (1980).

Advantages: Size distributions and analyses of pore-types obtained, by micromorphology and image analyses, cannot be derived from soil-physical measurements. In this way specific and unique information is obtained.

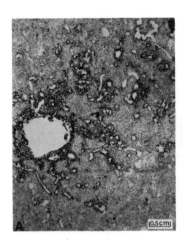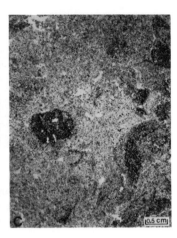

Fig. 7.2 Photographs taken of representative parts of horizontal thin sections made of stained soil samples at 37 cm depth in identical loamy typic Fluvaquents. A: pasture land with large faunal channels and smaller root channels. B: arable land with primary ploughpan showing root channels. C: deeply loosened arable land with secondary ploughpan containing dominantly (unstained) packing voids. All stained voids have dark-coloured walls. Quantification of the porosity is presented in Figure 7.3.

References
For describing thin sections:

Bullock P., Fedoroff, N., Jongerius, A., Stoops, G., Tursina, T. and Babel, U. 1985. Handbook for Soil Thin Section Description. Waine Research Publ., Wolverhampton, England.

Kooistra, M. J., 1985. Guidelines for standard abridged micromorphological descriptions of thin sections. Soil Survey Paper 16. Netherlands Soil Survey Institute, Wageningen, The Netherlands.

For Quantimet applications:

Ismail, 1975. Micromorphometric soil-porosity characterization by means of electro-optical image analysis (Quantimet 720). Soil Survey Paper 9. Netherlands Soil Survey Institute, Wageningen, The Netherlands.

Jongerius, A. 1973. Recent developments in soil micromorphometry. *In:* Rutherford, G. K. (Ed.), 1973. Soil microscopy, Proc. of the 4th Intern. Working Meeting on Soil Micromorphology p. 67-83.

Bullock, P. and Murphy, C. P. 1980. Towards the quantification of soil structure. Journal of Microscopy **120**, 317-328.

7.5 Submicroscopy

Principles: Submicroscopy includes all electronic techniques, other than light microscopy, that are applied to study undisturbed soil materials. Submicroscopy is applied when magnifications higher than $\times 200$ are required to study the structure and morphology of special morphological features and to study microchemical compositions of materials in situ. Two groups of techniques are distinguished: the non-destructive techniques and the destructive techniques. The non-destructive techniques include instruments in which the specimen is bombarded by electrons of high kinetic energy. With this process, X-ray photons are generated which contain analytical information about the specimen. Instruments included are the electron microprobe analyzer (EMA), the scanning electron microscope (SEM) with energy dispersive X-ray analyzer (EDXRA), the electron microscope microprobe analyzer (EMMA), the high voltage electron microscope (HVEM) and the scanning transmission electron microscope (STEM) combined with EDXRA.

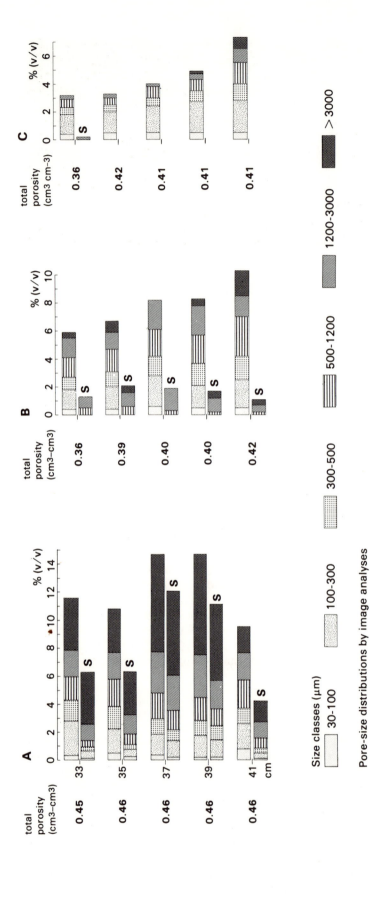

Fig. 7.3 Total porosities and pore-size distributions by image analysis with a Quantimet 720 from horizontal thin sections of stained soil columns. A: pasture land. B: arable land with primary ploughpan. C: deeply loosened arable land with secondary ploughpan (S = voids with stained walls).

X-ray photons can also be generated by bombardment with heavy particles like protons e.g. with particle induced X-ray emision (PIXE). With X-ray techniques most of the elements except the ultralight ones can be analysed. The lateral resolution is sometimes better than 1 μm and quantitative analysis is possible with a high degree of accuracy. Irradiation of a sample with X-rays can generate characteristic emission of electrons by the photo-electric effect. This technique, called electron spectroscopy for chemical analysis (ESCA), serves to identify the elements and the chemical bonding of the atom. The lateral resolution is not ideal as X-rays are difficult to focus.

The second group, of micro-analytical instruments are those in which some material of the specimen is eroded away from its surface either by ion bombardment or by irradiation with a laser beam. Mass separation of the secondary ions is accomplished by a mass spectrometer. The former category of instruments includes the ion microprobe mass analyzer (IMMA) and the secondary ion microscopy (SIM), the latter include the laser microprobe mass analyser (LAMMA).

The destructiveness of these techniques can be disadvantageous, but their sensitivity is 1 000 to 10 000 times better than with X-ray analytical techniques. The lateral resolution is in the order of 1 μm and all the elements of the periodic system can be analysed including hydrogen.

Application of these techniques requires specially prepared small-size thin sections which are highly polished and/or very thin.

Equipment needed: Submicroscopic techniques require very expensive, highly specialized equipment that is only available in some large research laboratories. Development of new, more advanced equipment proceeds rapidly.

Procedure: Detailed descriptions of the various procedures is beyond the scope of this text. The reader is referred to some recent summarizing articles on the topic.

Advantages: These methods allow observations of very fine structures at high magnifications and of the occurrence and distribution of chemical and mineralogical compounds *in situ*. Observations in situ of, for example, cementing compounds, which is possible with submicroscopy, may be very useful in studies of soil structure stability. The reader is referred to the references cited for case-studies.

Remarks: Submicroscopical analyses need to be calibrated with pure standards when performing microchemical analyses at high magnifications. For soils with their complex chemical composition, these standards are often not available. Progress is made rapidly, but results obtained are still sometimes difficult to interpret and quantify.

References:

Bisdom, E. B. A. (Ed.), 1981. Submicroscopy of Soils and Weathered Rocks. 1st Workshop of the International Working-Group on Submicroscopy of Undisturbed Soil Materials (IWGSUSM), Wageningen, 1980. Centre for Agricultural Publishing and Documentation (Pudoc), Wageningen, The Netherlands. 320 pp.

Bisdom, E. B. A. and Ducloux, J. (Eds), 1983. Submicroscopic Studies of Soils, Developments in Soil Science, Vol. 12, Elsevier, Amsterdam, 356 pp.

Bisdom, E. B. A., 1983. Submicroscopic examination of soils. Advances in Agronomy. Vol. 36 (1983): 55-96. Academic Press, New York.

7.6 Determining macropore-continuity

Principle: The vertical and horizontal continuity of large pores (macropores) is an important soil structure characteristic. It governs infiltration rates, particularly at higher rainfall intensities, and aeration processes, in clayey soils that have a soil matrix with fine pores. Macropore-continuity is difficult to study in a two-dimensional thin section, or in hand-specimen of soils in which only relatively small soil volumes are observed. Two approaches can be followed to derive macropore-continuity:

(1) Statistical studies, that translate two-dimensional images into a three-dimensional pattern. Recently, stereology has been applied in this context (e.g. Ringrose-Voase and Bullock, 1984).

(2) Functional characterization of soil macropores by staining techniques, or by using other tracers.

Equipment needed: Thin sections are used in detailed studies (see 7.3). Undisturbed blocks of soil can be percolated with water in which a dye has been dissolved. Also, a gypsum slurry can be used to fill macropores. The necessary equipment is minimal.

Procedure:

ad 1: The theory of stereology is mathematically complex. The reader is referred to literature (e.g. Weibel, 1979).

ad 2: Several procedures have been followed, depending on the specific objectives of soil structure studies. They include:

(1) **Infiltration patterns in dry clay soils with vertical cracks.**

Sprinkling irrigation results in the formation of small, 5 mm wide bands of water, along which water moves downwards following the vertical faces of the cracks. These bands are made visible by staining (e.g. Bouma, 1984).

(2) **Ponded infiltration in soils with vertical macropores.**

A thin gypsum slurry can be used instead of water. After hardening of the gypsum, observations can be made of the patterns of occurrence of the gypsum. Gypsum can only penetrate to a certain level through continuous pores. In turn, presence of gypsum indicates pore continuity (e.g. Bouma, 1984; Mackie et. al. 1984).

(3) **Formation of horizontal cracks upon drying of clay soils.**

Formation of horizontal air-filled cracks impedes the upward, unsaturated flow of water from the water-table to the rootzone. The percentage of horizontal air-filled cracks can be estimated, for a given pressure head, by using a staining test in undisturbed large blocks of soil that are encased in gypsum. They are then turned on their side. The upper and lower surfaces are opened and two sidewalls of the turned cube are closed. Methylene blue in water is poured into the cube and will stain the air-filled cracks. The surface area of these stained cracks is counted after returning the cube to its original position. A separate cube is needed for each (negative) pressure head. The K-curve for the peds is "reduced" for each pressure head measured in a cube. When, for example, 50% of the horizontal cross sectional area is stained, K_{unsat} for upward flow is 50% of the K_{unsat} at the same pressure head in the peds (e.g. Bouma, 1984).

(4) **Flow patterns in a wet clay soil.**

Undisturbed samples of a clay soil that have been close to saturation for a period of several months, are percolated with a 0.1% solution of methylene-blue or another dye that is absorbed by the clay particles. Thin sections are made, in which stained (continuous) and unstained (discontinous) macropores can be observed and quantified (e.g. Bouma, 1984). (See Fig. 7.2 and 7.3).

Advantages: Use of the above mentioned techniques allows the characterization of soil pores in terms of pore type, size, shape and continuity. Such characterizations are difficult to obtain with soil-physical methods, and they provide an opportunity to more specifically define soil structure.

References:

Stereology:

Ringrose-Voase, A., and P. Bullock. 1984. The measurement of soil structural parameters by image analysis. *In:* Bouma, J. and P. A. C. Raats (Eds.) 1984. Water and solute movement in heavy clay soils. Proc. of an ISSS Symposium. ILRI publ. 37. Wageningen, The Netherlands.

Weibel, E. R. 1979. Stereological Methods. Vol. 1. Practical Methods for biological Morphometry. Academic Press, London.

Pore continuity:

Various methods summarized by:

Bouma, J. 1984. Using soil morphology to develop measurement methods and simulation techniques for water movement in heavy clay soils.

Mackie, L. A., C. E. Mullins and E. A. FitzPatrick. 1984. Structural changes in two clay soils under contrasting systems of management.

Both articles appeared in the proc. of the Symp. on Water and solute movement in heavy clay soils (see complete reference under "stereology").

7.7 Use of soil morphology to calculate hydraulic conductivity

Principle: Pore sizes are directly measured in thin sections of undisturbed soil materials. Thin sections are appr. 20 μm thick plastic-impregnated slices through soil. Morphological observations allow distinction of different types of pores. Measured pore sizes are related to K using soil physical equations.

Equipment needed: Thin sections and microscope. Automatic image-analysis-equipment (e.g. Quantimet) is useful, but not essential. Calculations can be made with pocket calculators.

Procedure: Two procedures will be discussed. The first is intended for single grain soils (sands) in which soil pores consists of packing pores between the grains. The second is intended for well-structured clay soils with clearly developed aggregates, where measurements are focused on cracks between peds. Staining with methylene-blue is used to distinguish continuous pores.

Calculations: The first procedure is based on the equation of Marshall (1958):

$$K_i = \epsilon^2 n^{-2} \left\{ r_1^2 + 3 r_2^2 + 5 r_3^2 + \ldots (2n-1) r_n^2 \right\} / 8$$

where: K_i = intrinsic K (m^2). ϵ = porosity $(m^3 m^{-3})$; n = number of equal classes of the pore volume each with a specific size r. Here: $r_1 \, r_2 \, r_3 \, r_n$. r values are observed in thin sections. This method is not only used to obtain K_{sat} values but also K_{unsat} values.

The second procedure is based on flow through a plane slit with a defined width and hydraulic gradient (Bouma et. al. 1979):

$$K = \frac{\rho.g.d_n^3.l}{12 \eta A}$$

where: ρ = density of the water $(kg \, m^{-3})$
d_n = width of the "necks" in the cracks (m)
g = accelleration of gravity (ms^{-2})
l = length of the stained cracks in thin section (m)
η = viscosity of the water $(kg \, S^{-1/2} \, m^{-1/2})$
A = surface area being considered (m^2)

Values for d_n are obtained by a statistical procedure in which the width distribution and continuity of stained cracks is analysed (Bouma et. al. 1979).

Time required: Preparation of thin sections takes 4-6 weeks. Microscopic measurements can be made within a few hours. Use of electronic image-analysis equipment (Quantimet 720) is attractive. Calculations take little time.

Cost: Low when thin sections are available. High when thin sections have to be specially made.

Accuracy: Moderate, due to the very strong effect on K_{sat} of small changes in pore diameter. Accuracy in sands is as good as the accuracy obtained with calculations using moisture retention data. Values for clays were accurate, as calculated and measured in seven clay soils (Bouma et. al. 1979). Use of conceptual pore-interaction models reduces the accuracy of the method in general.

Advantages: The first procedure can be applied in thin sandy layers from which no physical samples can be taken. Application of the second procedure allows calculations in clays where other calculation models don't work.

Disadvantages: Only suitable for sands without much silt and clay (procedure 1) or clays with distinct, slowly permeable peds (procedure 2). Costly procedures if thin sections have to be made.

Remarks: These methods are particularly relevant for research purposes. Direct measurement of K_{sat} is cheaper and more rapid. However, procedure 2, in particular, is useful to better understand flow patterns and to explain differences observed. As stated, the first procedure can also be used to calculate K_{unsat}.

References:
Procedure 1 and 2:
Bouma, J. and J. L. Anderson. 1973. Relationships between soil structure characteristics and hydraulic conductivity. *In:* R. R. Bruce (ed.). Field Soil Moisture Regime. SSSA Special Publ. No. 5, Chapter 5, pp. 77-105.

Procedure 1:

Bouma, J. and J. L. Denning. 1974. A comparison of hydraulic conductivities calculated with morphometric and physical methods. Soil Sci. Soc. Amer. Proc. 38: 124-127.

Procedure 2:

Bouma, J., A. Jongerius and D. Schoonderbeek. 1979. Calculation of saturated hydraulic conductivity of some pedal clay soils using micromorphometric data. Soil Sci. Soc. Amer. J. 43: 261-264.